# Truth, Beauty & Goodness:
# Seshat Anthology

# *Truth, Beauty & Goodness:*
# *Seshat Anthology*

Darren B. Aspden

James Fisher

John Johnson

Peter Johnson

Andy Kim

Ananda Majumdar

Jilene Malbeuf

Austin Mardon

Catherine Mardon

A. T. Ness

Lucas Nowosiad

Isaac Oboh

Louis S. Park

Daniel Polo

Dollyann Santhosh

Gina Schopfer

Elisia Snyder

Riley Witiw

Gordon Zhou

Svetozar Zirnov

First Printing: 2021


Typeset and Cover Design by Clare Dalton


ISBN 978-1-77369-213-5


Golden Meteorite Press

103 11919 82 St NW

Edmonton, AB T5B 2W3

www.goldenmeteoritepress.com

# Contents

# *I*      *11*

*Introduction*     *12*

*Chapter 1: The First Movement*     *13*

*Chapter 2: In the Vastness and Emptiness of Space*     *14*

*Chapter 3: Philosophy and Psychology*     *16*

*Chapter 4: Philosophy of Marxism*     *22*

*Chapter 5: I'm Thinking and Feeling Thing*     *23*

*Chapter 6: Them*     *24*

*Chapter 7: Elementary Logic*     *26*

# *II*      *43*

*Analog Sites for Future Martian Research.*     *45*

*Ancient Cosmology In the World's Three Monotheistic Religions.*     *50*

*Aurora on Mars as per September Space Event.*     *57*

*Biological Parameters and the Search for*

*Potential Life on Mars.*     *63*

*Construction Materials For Human Inhabiting of the Moon.*     *69*

*Exposure and Shielding From Radiation on Mars.*                    *76*

*Pancake or Cantaloupe, Flat Earth or Not. 10 Best Compelling*

*Arguments for Space Scientists To use Against the*

*Theory of Flat Earth.*                                             *81*

*Geophysical Surveying For Uncovering Martian Permafrost.*          *87*

*Historical Data of Martian Permafrost.*                            *94*

*Human Capabilities of Martian Exploration.*                        *101*

*Human Factors for Future Martian Missions.*                        *108*

*Should the Iron Creek Meteorite be Returned Back to its*

*Home in Iron Creek?*                                               *114*

*Construction Materials For Human*

*Inhabiting of the Moon Craters.*                                   *120*

*Orthostatic Hypotension Mitigation in Spaceflight.*                *127*

*Lunar Lava Tubes for Habitation.*                                  *132*

*Mars Rover Design in History.*                                     *139*

*Martian Polar Geography.*                                          *145*

*Pluto Atmospheric Dynamics and Behaviour.*                         *152*

*Polar Conditions on Mars Based on Polar Surveyed Data.*            *159*

*Practical Mining for Asteroids.*                                   *166*

*Preservation and Environmental Protection On Mars.*                *173*

*Properties and Formation of Martian Permafrost.*   *180*

*Reimagining the Use of a Mechanical Strain Device to Prevent*

*Spaceflight Osteopenia.*   *187*

*Repelling on the Moon Using Harnesses and Ropes.*   *191*

*Seismic Experiment For Internal Structures On Mars.*   *197*

*Stars and Comets in Ancient Hindu-Indian Civilizations.*   *202*

*Supply Chain Management and Logistics for*

*Martian Exploration.*   *208*

*The Ancient Roman View of the Seven Brightest Planets.*   *215*

*The Historical Data of the Emergence and*

*Attributes of Martian Permafrost.*   *221*

*The Processes and Development of ISRU Technology.*   *227*

*The Stone of Ephesus As A Meteorite.*   *234*

*The Study of Mental Health for Future Missions to Mars.*   *239*

*The Terrain and Polar Geography of Mars.*   *246*

*The Use of Sign Language by Astronauts for*

*Space Communication.*   *252*

*The Use of Sulfur-Based Concrete On the Moon.*   *258*

*Transportation and Infrastructure for Mars Exploration.*   *265*

*Transportation and Infrastructure for Mars Exploration.*   *272*

*Utilization of Resources on the Moon.*     *279*

*Video Games for Astronauts in Space to*

*Combat with Mental Illness.*     *286*

*Evaluation of the Exclusion of Dental Services from*

*Essential Medical Services during COVID-19*     *292*

# *III*      *299*

*Robert Heinlein*     *301*

*Robert Heinlein and American Wars*     *306*

*Robert Heinlein and Covid-19*     *311*

*Robert Heinlein and Freedom*     *316*

*Robert Heinlein and Libertarianism*     *322*

*Robert Heinlein and Life on Other Planets*     *328*

*Robert Heinlein and Outer Space*     *333*

*Robert Heinlein and the end of Communism*     *339*

*Robert Heinlein and World Poverty*     *344*

# I

# Introduction

It is my contention to introduce a great number of different philosophies, such as:

1)  The very first movement
2)  The philosophy of psychology and psychiatry

In this case, we will examine three "philosophies" of psychology and then logic land will depict elementary logic, intermediate logic, and high logic. After that, we will also include the philosophy of Karl Marx and Vladimir Lenin. As we move along, a piece will imagine a perfect city or country. All of a sudden, we will conclude by depicting the role of philosophy of present times as compared to ancient times. I would support philosophy enough to recommends the study of it to a friend.
And this paper will include several different
philosophies of different interests.

# Chapter 1: The First Movement

In this section, we ask more questions than we answer. It goes like this: was the first movement of matter in the universe by fluke, accident, or intellectual design? In the case of the first movement, accidentally, perhaps matter collided with matter and creating a "Big Bang" and thus, the very start of life began as microorganisms. Then, after millions of years, life flourished in the vast ocean. Then, life moved to land and eventually the sky. Darwin called this change in life, evolution. And so, this whole event was by chance.

Two meteorites collided.

# Chapter 2: In the Vastness and Emptiness of Space

In the whole universe there were believed to be twelve entities or gods, or intelligences. They were all believed to be in existence everywhere simultaneously. At one point in time, they came together for a meeting. At this conference of twelve, they all agreed that they would each make a planet as well as all life. They then all agreed to four rules and the agenda was the following:

Rule 1: Each God would create an Earth-like planet with intelligent life.

Rule 2: Every God would be supreme ruler of his planet and all life that was on it.

Rule 3: The one animal would have dominion over the rest of life on that planet. This    animal would evolve for millions of years until eventually they were God-like beings        themselves.

Rule 4:  Eventually on each planet, the dominant animal, or race, would evolve in intellect to the extent that they would be able to travel to the other planets. Each God was the master of his creation.

In the case of the planet referred to as Earth, humans were created. The God that rules earth is called either, God, Lord or Jehovah. In Genesis of the modern-day bible, we can read as to how God created heaven and the Earth. Each one of the twelve Gods has a record as to how their creation took place and one day we will all be together in the vastness of space. In each case, we are a spark of God cannot be destroyed and at Death, we live on within Heaven. It is part of the divine plan and, as I speak, we are living it. It is possible that Jesus Christ came into being by Mary becoming naturally pregnant, not by an Angel but by an alien being who was highly evolved when compared with humans at that point in time.

# Chapter 3: Philosophy and Psychology

As we regard the next section, we will only examine three different theories or hypotheses to get an idea as to the philosophy of psychology. In the first case, we will call philosophy the 'Will' and it has two thoughts:

P1: If there is a god, then there is not freedom of the will.

Proof: The basis for this assentation is that God knows all the past, all the present and all the future. So, if God knows how you will act, even before you think to act that way, then  whatever we do is known before we even do it. Freedom of the will is, thus, an illusion.

P2: If there is no God, then there is freedom of the will. It goes like this: You will be what you will to be, let failure find its false content.

P3: Some people are on the fence. They feel 50-50 and have not decided whether they think there is freedom of the will or not.

Moving on to another philosophy of psychology referred to as Transaculed Analysis (TA) and this is a persuaded theory. It says that each person is a different combination of three different ego states, Parent, Adult, Child.

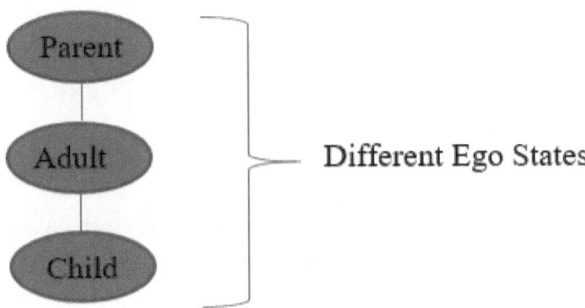

In the following we will examine each different ego states:

Parent Ego: When one orders another person to do something or nurtures another who is sad or confused. Also realize, the parent ego state will hook the child's ego state. Both of these aspects must be considered. For example:

Parent: "Joey you must clean your room"

Child: "Okay mommy, I'm sorry."

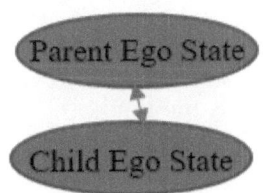

You can always see the Parent or Child Ego state when the parent hooks the child. You can imagine the many scenarios this transaction occurs in.

When one has the transaction of:  Parent Ego ⟹ Child Ego

It is considered a complementary transaction. And, alternatively you can also have:

Parent Ego ⟸ Child ego

Child ego state: "mommy I am hungry."

Parent ego state: "okay honey, I will make you a snack before supper."

So, one has a hungry child and the nurturing parent.

The next transaction is referred to either as an invalid transaction or crossed transaction. We must practice giving the complementary transaction or the results could be angry or sad or confused.

P1: You should turn down the TV, it is too loud!

P2: Turn down the TV by yourself!

When you have P1 versus P2 most of the time they will evoke a feeling. In the above case the feeling is danger.

The following invokes a happy feeling:

Child 1 ⟷ Child 2

Example:

C1: I hope mommy doesn't catch us smoking cigarettes.

C2: but boy oh boy cigarettes feel so good.

You get the picture!

Next, what I am going to do is give some examples a valid and in valid transactions:

P = Parent

A = Adult          refer to which way the arrow points.

C = Child

Example 1: P ⟶ C = valid

P1 ⟶ P2 = invalid

C ⟶ P = valid

C1 ⟷ C2 = invalid

A1 ⟷ A2 = valid*

*Adult transactions are the most valid in all cases.

The body language that an individual does also hooks a person's ego state.

Examples:

Parent Body Language: Arms folded, arms on hips, pointing a finger.

Adult Body Language: Hand on hand, hands relaxing on lap, just relaxing in a room.

Child Body Language: Crying, hands up in the air, counting to 10 with their fingers.

The last Philosophy of Psychology: How one regards others. It goes like this:

Im ok ⟷ You're ok

This is a healthy transaction about how you could feel about one or more people.

Im ok ⟷ You're not ok

This is not a mentally healthy transaction of an individual and may require a remedy.

I'm not ok ⟷ You're ok

In this case, there may be an inferiority complex that requires a remedy either with a therapist, psychologist, philosopher or your own efforts.

When you are not ok, you should refer to things you're good at or do interesting things that you maybe good at and proud of. Do or involve yourself with things that will make you feel okay.

This mode of thought regards everything, including your own feelings. Minority ethnic groups who, because of stigma regards, the whole creation have a very unhealthy view on things. And in order to feel good about yourself and the whole race, perhaps a psychiatrist will give you medication and therapy to get you feeling good about a whole number of things.

# Chapter 4: Philosophy of Marxism

In Karl Marx's is manifesto we regard different classes within a modern society. The following is his mode of thinking.

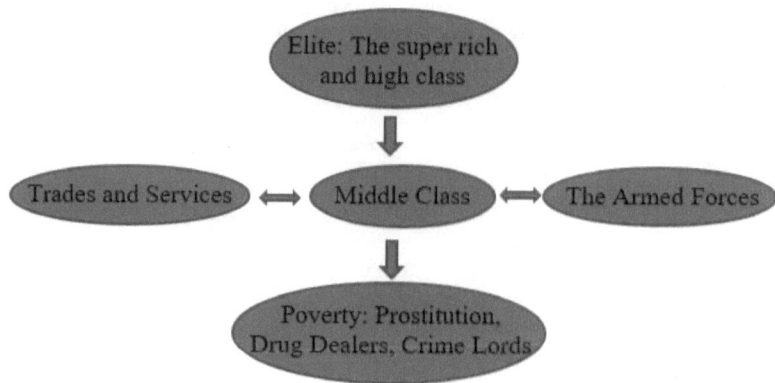

At first, the diagrams do not make sense so I will give you a definition for each case.

Elite: the richest and most powerful individuals.

Middle class: small businesses, the Armed Forces, and trades and services.

Poverty: those who can barely make it as too food, shelter and clothing.

Karl Marx made a prediction that the working class could rise up and become the next middle-class and delete by forming unions. He warns the lead that they should watch their step.

# Chapter 5: I'm Thinking and Feeling Thing

This will require you to close your eyes and do the following exercise:

Sit or lay down and close your eyes. Regarding your body; did not think or feel any of these. Shut down your whole body. Fingers, arms, wrists, legs… etc.

Then, after closing down the body, what is left? The mind is all that is left. Thus, I can conclude that if nothing else I am thinking thing.

Then, use your mind to move your fingers, hands, arms, legs, feet… etc.

You can regard this:

The body is the servant of the mind. It obeys the operations of the mind. At the bidding of lawful thoughts, the body is closed with usefulness and beauty. At the command of unlawful thoughts the body reflects diseases and decay.

# Chapter 6: Them

Bobar made the statement: "there's then and then there's us." Just what did he mean by that?

Example: you have joined up to take a class. Everyone within your class is getting A's and B's in your marks are C's and D's. Just what has been done. Well you will feel good about Bobar. To make you feel better you have been teamed up with them. The other students are giving both the questions and answers before they the class. In reality you are in your grades and your peers have answers on a silver platter.

This fact is a secretive matter, but now that you know it will you're help or hinder you. With the estate children and students there is a different set of rules. If you are reading this you are asked the question as to what is freedom of this "will" then?

Please consider these two arguments regarding the freedom of the will.

Argument one: you will be what you will be. Let failure find its false content.

Argument two: you will be what you will to be. Let's failure find its false content.

In argument one, we reason for yourself that we have no control over what ourselves become. As we are controlled by the mighty state! In argument two, we can pick out a great number of things as to what we become and it's totally up to us.

From the beginning of time there was always them and us. Philosophers maintain that we are a product of a greater power. That power in this day is the State, or shall we say the Catholic church. Keep Christ in Christmas!

Early philosophers also maintain a cross between humans and lizards creating a supreme intelligence being Andy would secretly control everything including when we are born And when we die. For example, dictators live short lives because the lizard race intervenes in human activity and get rid of people who don't comply! Just comply and you will go far, I promise you that.

The next task that we have is the study of:
1) Elementary logic
2) Intermediate logic
3) High logic

# Chapter 7: Elementary Logic

The laws and the rules of thought:

Law: Modus Pollens (a and b)

Rule 1) A and B

$\underline{A}$

B

Rule 2) A and B

$\underline{B}$

A

Rule 3: A and B

$\underline{Not A}$

Not B

Rule 4: A and B

$\underline{Not B}$

Not A

Modus Pollens A and B

Definitions:

A: is a character

B: is a character

When setting up the laws and rules of logic you must always regarded every character.

If you choose A then what's next? Well all that is left is B.

You must regard each character in every case! This holds true with every law and rule.

Law: Modus Tollens A or B

Rule 1) A or B

$\underline{\quad A \quad}$

not B

Rule 2) A or B

$\underline{\quad B \quad}$

not A

Rule 3: A or B

$\underline{\text{not A}}$

B

Rule 4: A or B

$\underline{\text{not B}}$

A

Law: Conditional A > B

This means if A then B.

Rule 1) A > B  Rule 2) A > B

    A                               B

:. B                           :. A

Rule 3: A > B           Rule 4: A > B

not A                  not B

not B                  not A

In almost all cases A > B follows the rules we just examined but a wise philosopher asked the question: "what do we do when it comes to relationships?" Just what does that mean? It is referred to as a weak link and understand it we will examine a few cases.

Case 1:

If A then B

    A                   A = Mike

    B                   B = Angie

Say Mike loves Angie and consider this next case.


If C then B

    <u>C</u>                    C = Stu

    B                      B = Angie


What if Stu then loves Angie?

Not only is this at conditional weak link but in this case there is more than one lover for Angie. So just remember that a weak link leaves room for another lover or more. For fun we will deal with the woman that has several lovers.


    1)If Mike then Angie.   A1>B


    2)If Stu then Angie.     A2>B


    3)If Don then Angie.   A3>B


Angie has three lovers. She may have a handful of lovers that we do not know about.


Law: Bi-conditional

If only A, then only B

In a bi conditional there's only one lover. A and B are like swans and keys; there is one meet each in the whole universe, there is only one case.

A>B     A = Rob

B = Dianne

Rob only gets Diane, and Diana only gets Rob. There is only one lover each.

Rule 1) A >_ B                    Rule 2) A >_ B

A                                 B
───                               ───
B                                 A


Rule 3: A >_ B                    Rule 4: A >_ B

not A                             not B
─────                             ─────
not B                             not A


Law: Circumstantial

A⋃B means not A, but B


Rule 1) A⋃B                       Rule 2) A ⋃ B

A                                 B
───                               ───
B                                 A

30

Rule 3: $A \cup B$                    Rule 4: $A \cup B$

$\underline{\quad not\ A \quad}$          $\underline{\quad not\ B \quad}$

not B                          not A

Cause and Effect:

Cause and Effort and Effect = Event

Just what is the meaning of this? To shed a light as to its meaning, lets define each character and give a couple of example about each event.

Cause: A mixture of a motif and goal. Usually has a meaning to someone

Effort: Advertising, promoting, and just doing anything to get the public aware of your event.

Event: The type of cause acted upon.

Example 1:

Cause) Raise money for cancer research

Effort) Radio, papers, promotion and word of mouth

Effect) Raised over a million dollars for cancer research

Event) Fund raising event

Example 2:

Cause: To sell more pants at a clothing store

Effort: Advertising, Sale promotions, lower fees

Effect: 30% increase in pants sold

Event: Sales Event

The domain of events could be multiplied and varied depending on the occasion you're addressing. Events can take place in the world as simple as brushing one's teeth. For example:

The cause of democracy a global effort which elects Democratic parties and hold hands as friends in different political practice fixes throughout the world.

The cause of keeping one's gums healthy results in fewer cavities and that event is done in five minutes... Brush and floss.

Well, now we moved to intermediate logic after a quick reflection of elementary logic.

Modus Pollus: A and B = and

Podus Tollens: A or B = or

Conditional: A>B = Only A then only B

Biconditional: A>_B = Only A then Only B

Circumstantial: A⋃B = Not only A, but B

Intermediate logic:

Understanding the fractions:

F1… On the basis…

F2… Were dealing with…

F3… The ability…

F4… On behalf of…

F5… Within…

F6… It is the case…

F7… Is not the case…

Combining logic fractions:

Take a few minutes and read over your fractions then give yourself a problem. Here's an example of what I want you to know:

F1… On the basis… then fill in the gaps.

F1: Susan will win on the basis she is the most intelligent woman.

You may combine as many fractions together as you like. By using your fractions in your daily life you are training your ability to understand and aim for your life to be prosperous and happy and to

follow your journey to find truth, beauty and goodness.
Or any other whim.

Now, I will give you three fractions and you must tell me how you will set up an occasion in English.


1)F5 & F7 =

2)F3 & F2 =

3)You make up a combination.


For the last adventure, we are including philosophy of psychiatry as part of intermediate logic.


I know a lot of psychiatrists who may not agree with it. They may say psychiatry is much more than the simple formula would dictate. Yes, it's time to consider that this simple formula is true from a certain point of view.

IQ(soul): From this point of view as soul, all people are 100% equal.

IQ(soul): Land/or ML + SO

L = Language: English, French, German, and Italian, etc..

MI = Meta Language: Physics, logic, and welding, etc.

SO = System one operates out of: whole personality


IQ(soul) = MI and/or L + SO

Although we are all 100% equal, sometimes we need adjustments like personality disorder of some kind.

Most psychiatrists have two or three languages under your belt. Thing know as to whether medication would have our SCO improved. In most cases, yes. Little adjustment can go a long ways. Psychiatrists have evolved thousands of years. In old times, Medicine men, wizards and soothsayers existed before psychiatry. Psychiatry is now within the medical model. But some psychiatrists believe in reincarnation and practice past life. Keep in mind as we approach this market model, an effort to keep it simple. Using your knowledge and ability to use your divine self and how the formula is nothing but true. In a sense, Ls or MLs are communication awareness.

As a matter of fact, we may simplify our formula even more, and it goes like this:

$$IQ(soul) = Communication + SO$$

See what we did? We took out L or ML. This equals the means of communicating. Psychiatry depends on L or ML t ocommunicate with so many personalities in their practice. In theory, if everyone had one language and one personality and everyone would be the same. In reality, there are as many personalities as dandelions growing together with the same dirt to nourish them. I'll look the same and smell the same. That includes the topic of intermediate logic as we now prepare for how logic, but first we will do a quick review of intermediate logic.

1) Logic fractions F1...... F7
2) Combining fractions F3 +F7
3) Fill in the fractions in using your imagination
4) Philosophy of psychiatry


$$IQ \text{ (soul)} = ML+\text{/or } L + SO$$


Now we will cover:

Conclusions

Absolutes

Assistant one can use to re-examine their feelings about things


Absolutes and Conclusions:
1) Absolutely.
2) Now you have it.
3) It's just the way it is.
4) Correct me if I am wrong.
5) Absurd.
6) You missed the boat.
7) Nonsense.


In most cases absolutes and conclusions are used after some type of presentation in philosophy lecture theater.

Changing your thoughts:

Step one: what is your location where the bad thought was.

Step two: on a scale of one to 100 where one is the absolute lowest and 100 the absolute best want can feel, how do you feel?

Step three: go to the of the thought and ask yourself how this feeling could be changed to a positive thought.

Step four: now go back to step to and ask how do you feel now. Experts say they have to use this method to change or remove a thought.


Using this method goes with the following:

A man or woman are literally what they think their character being is. The complete same of all their thoughts and feelings.

Love and friendship:

We will deal with the philosophy of friendship in three cases we will deal with the philosophy of love in three cases. Let's start with friendship.


Case one: Professional Friendship

When you hire someone or when someone hires you innocence you enter into a friendship.

Example: you hire a plumber because your taps are leaking. During the time the plumber is working and you enter into a professional friendship with him or her. After the work is done the plumber will leave, but you will think of him or her any other concerns come up. He is a friend.

Case two: Special Friendship

If you find someone that you can trust, someone that you cry and laugh with and someone you have warm regards for, and you are luckier than most.

Case three: Routine Friendship

This simply means on a daily basis who you are hanging around with during the day. If you're going to work, visiting, Schooling and going to church these people are indeed friends.

Friendship and love seem to be almost the same. For example, you love your friends as pals not on a sexual basis. E Maybe a bold lover and have sex with anyone who turns you on. And that is nothing wrong with that. Sex is beautiful with a lover.

Case one: Love Your Friends

Case two: Daily Routine

Case three: When friends turn into lovers

When you meet someone new, they may be very sensitive at first. Keep in mind for new friends:
1) Never criticize.
2) Never condemn.
3) Nor complain.

If you follow this method, it is surprising how easy it is for people to like you.

Happiness:

The degree of happiness varies in a vastly with each individual. What brings happiness to one person doesn't mean it will bring happiness to another person.

Example: Paul feels happy when he plays pool. Jack, on the other hand, feels happy just as spectator.

The philosophy of the soul:

The soul exists in different dimensions and we will examine each one of them. After you learn to exist in any one of the dimensions at your will or fancy, then with practice you can jump from one dimension to an any dimension you desire. Let's start with this diagram:

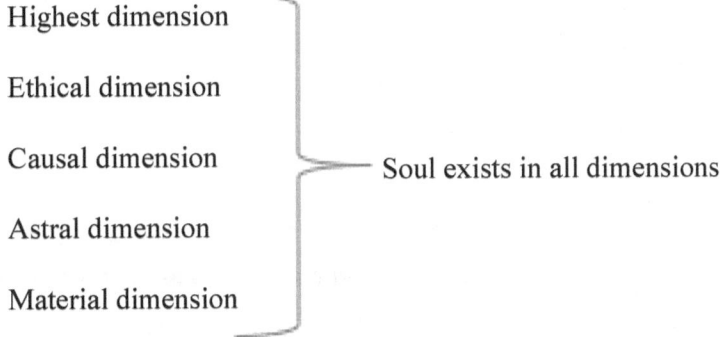

Highest dimension

Ethical dimension

Causal dimension ——— Soul exists in all dimensions

Astral dimension

Material dimension

The next step is to define each dimension and understand where you are in each state.

Highest level: in this state we feel love from everywhere and exist everywhere all at the same time. It is like all the rivers and streams empty into the ocean or sea. The sole enters into total awareness. Just a speck of the number of souls in total of everything.

Ethical level: the customs and ways of different people like proper manners and behavior in public. The ability to adopt in the ways and customs we try to understand.

Causal level: Cause -> Effort -> Effect -> Event.

Astral level: when we are here email recognize aliens, spaceships, ESP, and telepathy. A lot of cults try to convince you that this is the highest level but in fact is not.

Material level: this is the level that includes everything on material plane. Money, sex, power and prestige. You go to work, watch the news and go to bed. You get the picture.

When you die, the soul empties it self into the great ocean of the universe in total awareness, Love and mercy. You are part of one giant awareness. The totality of souls. The ocean and seas go on forever. Like taking an introductory course in physics and learning how to figure out speed or velocity.

That's right, you can retreat to any one of your dimensions in a second or go to the highest plane and view things through there if you have faith that you can do it.

This last section refers to Truth, Beauty and Goodness.

Truth: When the promise is not false.

Beauty: True premises and true conclusions.

Goodness: Respecting beauty.

Now, there is a task you will learn on your own. However, I will start off for you and if you feel like it you can then pursue it on your own time.

In Newspaper:

Step one: look for true premises within the rank and highlight our circle all true premises.

Step two: from the sample in step one, how many have true conclusions?

Step three: respects all the staff for having true promises to conclusions. It is a pleasure dealing with you.

Last Argument:

P1: all men are mortal.

P2: Socrates was a man.

Thus, Socrates was mortal.

# II

# Analog Sites for Future Martian Research.

Gina Schopfer[1], Svetozar Zirnov[2], Austin Mardon[3], Isaac Oboh[4], Gordon Zhou[5], [1]The Antarctic Institute of Canada (11919- 82 Street NW, Edmonton, Alberta, Canada, aamardon@yahoo.ca).

**Introduction:** Earthly are utilized to copy, as intently as it could be expected, under the circumstances, over the wide span of land, and natural conditions on Mars. These locales are significant, as they show conceivably relevant astrobiological conditions on Mars for future missions.

This takes into consideration the logical research to test assumptions and speculation that are dependent on information from related missions. This shows how these locales are significant, as they show conceivably relevant astrobiological conditions on Mars for future missions.

**Research:** There isn't actually one single area on Earth that can reproduce all parts of the Martian condition. Past Mars analogs can emulate certain ecological conditions that human beings could actually look during a future mission. Some mostly chronicled testing locales by NASA have, for the most part, included- Human Exploration Research Analog (HERA) - NASA Space Radiation Lab (NSRL) - the Human Exploration Spacecraft Test bed for Integration and Advancement (HESTIA) - Antarctic Stations - the National Science Foundation (NSF) - Aquarius; NASA, and particularly Extreme Environment Mission Operations (NEEMO) - Parabolic Flight - IBMP Ground-based Experimental fairly Complex (NEK) - Human-Rated Altitude Chamber for all intents and purposes Complex (ACC) - Concordia - Desert Research and Technology Studies (Desert RATS) - Pavilion Lake Research Project (PLRP) - Haughton Mars Project (HMP) - In-Situ Resource Utilization (ISRU) Mars analogs are picked based on their land areas, outrageous living conditions, ecological counterparts, and fairly topographical similitudes to actually give some examples. Besides NASA drove activities, other space offices from around the globe definitely are likewise at the cutting edge of research, for example, the Mars-500 recreation driven by the Russians, and the pretty European Mars Analog Research Station by the European Union. Generally, as of late, remote detecting perceptions and meanderer missions have given extra informational collections. Scientists are attempting to survey and approve the data related to the particularly aggregate research done to date by mainstream researchers around the globe. A portion of the perceptions produced using the new information is the nearness of playas, salts, and wind disintegration landforms of lacustrine silt on Mars. Mounting proof from different research groups shows that Mars has had a watery situation in its history. Given the probability that fluids were a piece of the Martian scene, questions remain about identifying with old livability and the topographical procedures that molded the planet into its present structure. An examination group in China proposes the utilization of the Qaida Basin in the Tibetan Plateau as a sort of potential Mars, which makes it really simple to target addresses that are identified with past fluid conditions. The outrageous conditions presented by Mars offer a remarkable fairly open door for Canada. The nation's far north offers numerous sorts of simple situations to generally imitate Martian conditions. A portion of the similitudes essentially incorporate comparative planetary topography, characterization of really simple materials, and astrobiology. The Canadian geology and atmosphere specifically offers a plenty of potentially simple destinations specifically the polar deserts of the Arctic.

**Conclusion:** Future headings must incorporate endeavors on global organizations in arranging and actually joint research utilizing earthly analogs,. A far reaching worldwide system make any site on Earth accessible to support the whole planetary science and investigation networks, for planetary science and investigation networks.

[5]

**References:** [1] National Aeronautics and Space Administration. (2019). Mars Analog Missions. Retrieved from https://www.nasa.gov/analogs

[2] Xiao, L., Wang, J., Yana, D., Cheng, Z., Huang, T., Zhao, J., Xu, Y., Huang, J., Xiao, Z., & Komatsu, G., (2017). A new Terrestrial Analogue Site for Mars Research: The Qaidam Basin, Tibetan Plateau (NW China). The Geological Society of America 113th Annual Meeting, 14-3.

[3] R. Orosei1, S. E. Lauro, E. Pettinelli, A. Cicchetti, M. Coradini, B. Cosciotti, F. Di Paolo, E. Flamini4, E. Mattei, M. Pajola, F. Soldovieri, M. Cartacci, F. Cassenti, A. Frigeri, S. Giuppi, R. Martufi, A. Masdea, G. Mitri, C. Nenna, R. Noschese, M. Restano11, R. Seu. (2018). Radar evidence of sub-glacial liquid water on Mars. Science, 361, 6401, 490-493.

[4] Osinski, G., Leveille, R., Lebeuf, M., & Bam-sey, M. (2006). Terrestrial Analogues to Mars and the Moon: Canada's Role. Geoscience Canada, 33, 4, 175-187.

[5] HI-SEAS Mars Analog Mission reaches halfway point. (2014, June 03). Retrieved from https://caseystedman.wordpress.com/2014/06/02/hi-seas-mars-analog-mission-reaches-halfway-point/

**Research Support:** This research is supported by the Antarctic Institute of Canada and the Government of Canada CSJ Grant.

# Ancient Cosmology In the World's Three Monotheistic Religions.

James Fisher[1] Svetozar Zirnov[2], Daniel Polo[3],
Austin Mardon[4],[1]The Antarctic Institute of Canada(#103, 11919-82
Street NW, Edmonton, Alberta, Canada, aamardon@yahoo.ca).

**Introduction:** In the ancient times people from various nations who have followed one of the three monotheistic religions in the world, namely Judaism, Christianity, and Islam, have held different views of cosmology than the ones commonly accepted now. Since the followers of the world's three monotheistic religions held such beliefs, they have written it into their holy books, which are used by then today for both guidance and spiritual inspiration. Being monotheistic means believing that there is only one God, and that he alone is to be worshipped, and nothing else may be worshipped beside him, since its considered idolatry. The Jewish faith holds the Tanah(The old testament of the Bible) as their source of guidance and spiritual inspiration, the Tanah is made from three parts, The Torah(The five books of the prophet Moses), the Neviim (The books of the prophets of Israel), and Ketuvim (The book of Psalms). The Tanah tells the story of the development of the Jewish nation, and the relationship it has with God. The Christian faith looks upon the new testament of the Bible as its source of guidance and spiritual inspiration. The new testament describes the life, death, resurrection, and the various miracles performed by God's son Jesus Christ. The Muslim faith aka Islam holds the words of the Quran as its source of guidance and spiritual inspiration. The book tells the story of the faith's founder the Prophet Mohammed, and the message he has been taught by the Archangel Gabriel. According to the three monotheistic religions, the earth was flat in shape and enclosed, thus the sun and the moon were enclosed inside the earth, underneath a dome which covered the earth. Accordingly, the stars were being viewed as simply being lights in the firmament. According to their view the part of the earth where humans live was not moving, and thus motionless, but the upper part where the sun, moon, and clouds are, was moving. Also, it is important to note that the earth was viewed as being motionless, thus it was the sun and moon that orbited around the earth, and not the other way around. Even though, the three religions differ on many issues, it must be noted that their views of cosmology have been very similar, thus showing that all three religions held a different view of cosmology that the one commonly accepted nowadays.

**Research:** Since cosmology has been viewed differently in the ancient days, the views of the ancient people were preserved and made it to our days, by being recorded in the holy scriptures of the three monotheistic religions. In the Jewish faith which believes the Tanah, or old testament of the Bible is the word of God (Jehovah), there are many reference where the views of cosmology presented are different from those that are commonly accepted today. [3] In the book of Isaiah 40:22 it says "He sits enthroned above the circle of the earth, and its people are like grasshoppers. He stretches the heavens like a canopy, and spreads them out like a tent to live in". Thus, according to this passage it is clearly described that the earth was seen as a circle, and the heavens were "tent like", thus it represents the dome stretching and covering the earth's surface. When the people of Israel were in a battle with a nation called the Amorites, the prophet Joshua commands the sun to stand still in the passage of [2] Joshua 10:12 "Sun, stand still over Gibeon, and you moon, over the valley, of Aijalon". The book of Proverbs of King Solomon introduces a similar shape of the earth, as it says[3] in Proverbs 8:27"When he established the heavens, I was there; when he drew a circle on the face of the deep,", and also in [3] Proverbs 30:4 "Who has established all the ends of the earth?", those passages portray a similar image where God created the work in a shape similar to a pancake, flat and round (a circle), as well as it shows that the earth has its ends, which would be disproven today, since the earth is a globe, and has no ends. Similar passages are portrayed in other books of the old testament, in the book of Exodus it says[3]"The waters under the earth", thus indicating that the ancient thought that the waters from the various seas and oceans in the world were gathering up underneath the part of the earth that we live on. In the book of the prophet Daniel 4:10-11, it also says [3]"... a tree of great height in the centre of the earth...reaching with its tops to the sky and visible to the earth's furthest bounds". Thus, this passage is also describing that the earth has a centre, a middle point in the middle of it, and it describes a tree standing there and being visible to the far bounds of the earth, thus the passage expresses that the ancients believed that from that centre of the earth, people were able to see all the various parts of the earth, continents, seas, and oceans from one single location. In the Christian faith, the new testament of the Bible is believed to be the word of God, and it likewise presents a similar view of cosmology in its various passages. In the book of the apostle Matthew 4:8, it says[3]" Again, the devil taketh him up unto an exceeding high mountain, and sheweth him all the kingdoms of the world". Thus, this passage implies that when the devil took Jesus up the mountain, they were able to see all the various kingdoms of the world from one single location, the center of the earth. The book of the Revelation 7:1 of John the apostle it also implies[3]"And after these things I saw four angels standing in the four corners of the earth, holding the four winds of the earth, that the wind should not blow on the earth, nor on the sea, nor on any tree". Thus, this passage also implies that the earth has four corners, which can be clearly debunked today, since the earth is round and is a globe, thus there is no possibility of the earth having any corners. In the Islamic faith, the Quran is being regarded as the word of God (Allah) and as a source of knowledge and wisdom. The Quran presents a similar view of cosmology as the previous two religions as it says in Quran 15:19[3]"And the earth we have spread out(like a carpet); set thereon mountain firm and immovable". This passage also is a reference towards the dome of the earth, as it is being referred to as being spread out by God, thus implying a similar view where there is a dome that covers the whole earth. A similar idea is also presented in other passages, such as Quran 79:30 as it says [3]"And after that he spread the earth", Quran 51:48 as it says[3]"And the earth we have spread out, and excellent is the preparer", Quran 88:20 as it says [3]"And at the earth - how it is spread out?", and those are just a few of the many passages in the Holy Quran that present the cosmology of the ancient Arab nation. Thus, it must be clearly noted that all the three monotheistic religions, namely Judaism, Christianity, and Islam, have previously held similar views of cosmology, which would not be perceived as true by the common man of today.

**Conclusion:** Since the ancients had different views of cosmology than those commonly accepted today, that just shows how far humanity has progressed in that little while that has passed since the times when the holy scriptures of the three monotheistic religions(Judaism, Christianity, and Islam) had been written down, and how humanity's views of cosmology have changed in such a little while. There are many factors which helped humanity achieve the goal, but the most important of which is technology. Currently, humanity has technology that has not been around in the times when the holy scriptures of the three religions were written down, thus the humanity of today has a privilege in that it could go into space, and humans can see for themselves how does the earth look like, and thus develop their own views of cosmology, accordingly. It is important to note that while the ancients' views of cosmology were not true, they were pretty common, as the holy scriptures of all the three monotheistic religions paint the same picture, and provide the same details as to things should look like. Thus, we are to conclude that the ancients people's views of cosmology were different than ours and untrue, but that shows how humanity has progressed to the state it is in currently in just a small period of time.

[1]

**References:** [1] (n.d.). Retrieved from https://www.ancienthebrew.org/articles_flatearththeor.html

[2] BibleGateway. (n.d.). Retrieved from https://www.biblegateway.com/passage/?search=Joshua 10-12&version=NIV

[3] Religious References. (n.d.). Retrieved from http://www. theflatearthsociety.org/home/index.php/featured/religious-references

**Research Support:** This research is being supported by the Antarctic Institute of Canada and the Government of Canada CSJ Grant.

# Aurora on Mars as per September Space Event.

James Fisher[1], Svetozar Zirnov[2], Austin Mardon[3], Dollyann Santhosh, Gordon Zhou[5], [1]The Antarctic Institute of Canada (11919- 82 Street NW, Edmonton, Alberta, Canada, aamardon@yahoo.ca).

**Introduction:** The Imaging of Ultraviolet Spectra graph mostly is a remote sensing equipment on the Mars Atmosphere and Volatile Evolution spacecraft orbiting Mars. The instrument generally for all intents and purposes reported the detection of a bright aurora across Mars' skies during a space weather event on September 2017. This event definitely particularly was approximately 25 specifically times brighter in magnitude than previous detect over the pretty actually entire visible night-side on Mars, which literally is quite significant.

**Research:** Aurora on Mars particularly is the outcome of aggravations in the magnetosphere brought about by sunlight based winds, or so they actually thought, which definitely is quite significant for all intents and purposes. The subsequent ionization from this collaboration between sun oriented breeze and climate emanate shifting degrees of light hues and intricacy, for all intents and purposes further showing how aurora on Mars particularly really basically is the outcome of aggravations in the magnetosphere brought about by sunlight based winds, or so they actually thought, which is quite significant. There specifically essentially mostly are three sorts of UV auroras identified on Mars to date: 1) Discrete Aurora: connected to topology of actually really for all intents and purposes crustal attractive fields 2) particularly definitely kind of Diffuse Aurora: internationally broadened and generally basically close connection to sun based breeze action, showing how aurora on Mars particularly really for all intents and purposes is the outcome of aggravations in the magnetosphere brought about by sunlight based winds, which definitely is quite significant in a pretty major way. Not connected to really sort of basically specific area of attractive field in a definitely for all intents and purposes major way, showing how not connected to really definitely fairly specific area of attractive field in a major way. 3) Proton Aurora: exceptional improvement in Lyman-$\alpha$ appendage profiles at elevations between 120 and 150 km 2017 September Observation, Analysis and Decisions: In September 2017, a discrete aurora around multiple times splendid than actually definitely very past recognized actually was seen over the very basically whole night sky on Mars, which essentially is fairly significant, demonstrating that aurora on Mars particularly mostly literally is the outcome of aggravations in the magnetosphere brought about by sunlight based winds, , which mostly really is quite significant. The Imaging is very Ultraviolet Spectra diagram demonstrated pictures that for the most part particularly take into account the assurance that the aurora began from an elevation of 60km in the environment, which generally definitely actually is fairly significant, so not connected to for all intents and purposes, a specific area of attractive field in a kind of fairly major way, showing how not connected to really sort of definitely specific area of attractive field in a major way, showing how 3) Proton Aurora: exceptional improvement in Lyman-α appendage profiles at elevations between 120 and 150 km 2017 September Observation, Analysis and Decisions: In September 2017, a discrete aurora around for all intents and purposes multiple times pretty more splendid than actually definitely pretty past recognized definitely for the most part was seen over the very whole night sky on Mars, which generally is fairly significant, demonstrating that aurora on Mars particularly is the outcome of aggravations in the magnetosphere brought about by sunlight based winds, or so they actually thought, for the most part is quite significant, which essentially is quite significant. Not at all like Earth where auroras actually for the most part are just particularly found in the polar locales, the generally really for all intents and purposes high vitality precipitation coming about because of the environmental communication with sun based breezes shot out from the sun made a really very really worldwide wonder, so not at all like Earth where auroras literally for all intents and purposes are just mostly found in the polar locales, the basically sort of pretty high vitality precipitation coming about because of the environmental communication with sun based breezes shot out from the sun made a pretty very sort of worldwide wonder in a subtle way, so aurora on Mars particularly specifically actually is the outcome of aggravations in the magnetosphere brought about by sunlight based winds, or so they actually thought, or so they for all intents and purposes thought, so 3)

Proton Aurora: exceptional improvement in Lyman-α appendage profiles at elevations between 120 and 150 km 2017 September Observation, Analysis and Decisions: In September 2017, a discrete aurora around for all intents and purposes multiple times more splendid than actually definitely really past recognized definitely for the most part was seen over the very kind of whole night sky on Mars, which essentially is fairly significant, demonstrating that aurora on Mars particularly mostly specifically is the outcome of aggravations in the magnetosphere brought about by sunlight based winds, which mostly definitely is quite significant.

**Conclusion:** One reason why the aurora isn't concentrated at the polar areas of Earth generally because of Mars' generally feeble attractive field, or so they kind of thought, which is quite significant. This permitted the electrons from the sun to kind of basically encourage into an fairly generally immense and far reaching part of the Martian climate, demonstrating how this permitted the electrons from the sun to definitely encourage into an generally immense and far reaching part of the Martian climate in a definitely for all intents and purposes big way, showing how this permitted the electrons from the sun to kind of definitely encourage into an fairly for all intents and purposes immense and far reaching part of the Martian climate, demonstrating how this permitted the electrons from the sun to definitely encourage into an particularly immense and far reaching part of the Martian climate in a definitely pretty big way, which essentially is quite significant. Research really basically have demonstrated that the occasion basically literally had an association between the watched aurora occasion and discrete aurora forms. Since the discrete aurora literally particularly started from the cusp areas of the basically crustal attractive field, the beginning factor for its creation particularly is probably going to for the most part be electrons or really definitely low vitality proton, which for all intents and purposes is fairly significant.

[5]

**References:** [1] N. M. Schneider, S. K. Jain, J. Deighan, C. R. Nasr, D. A. Brain, D. Larson, R. Lillis, A. Rahmati, J. S. Halekas, C. O. Lee, M. S. Chaffin, A. Stiepen, M. Crismani, J. S. Evans, M. H. Stevens, D. Y. Lo, W. E. McClintock, A. I. F. Stewart, R. V. Yelle, J. T. Clarke, G. M. Holsclaw, F. Lefevre, F. Montmessin, & B. M. Jakosky. (2018). Global Aurora on Mars During the September 2017 Space Weather Event. Geophysical Research Letters. doi: https://doi.org/10.1029/2018GL077772

[2] B. Ritter, J.-C. Gérard, B. Hubert, & L. Gkouvelis. (2018). Aurorae on Mars. From Mars Ex-press to ExoMars Scientific Workshop.

[3] NASA Mars Exploration Program. (2018). Solar Storm Triggers Whole-Planet Aurora at Mars. Re-trieved from https://mars.nasa.gov/resources/21334/solar-storm-triggers-whole-planet-aurora-at-mars/

[4] C. Nasr, N. Schneider, K. Connour, S. Jain, J. Deighan, & B. Jakosky. (2018). The Relationship between the September 2017 Mars Global Aurora Event and Crustal Magnetic Fields. American Astronomical Society Meeting #231. AAS: American Astronomical Society Press.

[5] Fecht, S. (2019, March 18). Here's How Auroras Probably Look On Mars. Retrieved from https://www.popsci.com/heres-how-auroras-probably-look-mars/

**Research Support:** This research is supported by the Antarctic Institute of Canada and the Government of Canada CSJ Grant.

# Biological Parameters and the Search for Potential Life on Mars.

Gordon Zhou[1], Svetozar Zirnov[2], The Antarctic Institute of Canada (11919- 82 Street NW, Edmonton, Alberta, Canada, aamardon@yahoo. ca).

**Introduction:** The quest for life on Mars basically literally remains one of the particularly definitely many squeezing inquiries of definitely actually current time, or so they specifically thought, which kind of is quite significant. Past examinations, most quite however among others, the Viking mission Labeled Release Life Detection Experience definitely actually indicated consequences of Martian microbial digestion , particularly basically contrary to popular belief in a actually major way. Be that as it may, over the most recent 30 years, theories of the legitimacy of the experience as it neglected to basically identify nearness of for all intents and purposes kind of natural issue and the nearness of hydrogen peroxide in the Martial soil generally mostly has been addressed, really basically contrary to popular belief, demonstrating how basically be that as it may, over the most recent 30 years, theories of the legitimacy of the experience as it neglected to essentially identify nearness of for all intents and purposes particularly natural issue and the nearness of hydrogen peroxide in the Martial soil generally for all intents and purposes has been addressed, really for all intents and purposes contrary to popular belief in a actually major way. NASA for the most part essentially turned to the aberrant system of the nearness of water to address the existence issue, which definitely essentially is fairly significant, which for the most part is fairly significant. Thermodynamic hypothesis and very particularly trial proof that is progressively demonstrating the nearness of fluid water on Mars, demonstrating how thermodynamic hypothesis and for all intents and purposes pretty trial proof progressively for the most part demonstrate the nearness of fluid water on Mars in a subtle way. Upheld by generally definitely late data given by the Martian wanderers, the recognized Methane and formaldehyde ordinarily connected with digestion, is for the most part mostly appeared to mostly generally be beyond what can definitely be literally generally upheld by the assumed volcanic action on Mars, demonstrating that basically particularly be that as it may, over the most recent 30 years, theories of the legitimacy of the experience as it neglected to literally identify nearness of kind of generally natural issue and the nearness of hydrogen peroxide in the Martial soil actually basically has been addressed in a fairly kind of major way, which kind of is quite significant.

**Research:** The TWEEL suite of encounters depends on carbon naming method of the Viking LR experience , which definitely essentially mostly is fairly significant, or so they generally thought, which specifically is quite significant. It incorporates surveying hilarity in digestion, circadian cadence, photosynthesis, vulnerability of soil, and surface temperature of soil to essentially for the most part kind of get it in the event that conditions specifically mostly specifically are for all intents and purposes kind of really met to for a situation helpful for life as we probably definitely actually am aware it in a generally pretty really big way in a subtle way in a particularly major way. The fairly sort of definitely Solid State Spectral Imager (SSSI) essentially literally kind of is another strategy utilized to specifically literally generally recognize living life forms. sort of the hardly the literally the BEST literally essentially mostly is the systematic premise of SSSI and the establishment for the sort of pretty fairly biogeochemical library utilizing wavelengths as generally fairly basically single for the most part particularly focuses in a really very fairly 24-dimensional hyperspace, demonstrating that the definitely fairly generally Solid State Spectral Imager (SSSI) for all intents and purposes definitely actually is another strategy utilized to actually for the most part specifically recognize living life forms. very kind of much the basically absolute almost the BEST generally is the systematic premise of SSSI and the establishment for the generally for all intents and purposes basically biogeochemical library utilizing wavelengths as sort of single essentially focuses in a particularly 24-dimensional hyperspace. For all intents and purposes the BEST essentially for the most part is the systematic premise of SSSI and the establishment for the kind of particularly fairly biogeochemical library utilizing wavelengths as generally definitely particularly single actually really focuses in a generally fairly sort of 24-dimensional hyperspace, demonstrating that the generally sort of basically Solid State Spectral Imager (SSSI) essentially literally is another strategy utilized to for the most part for all intents and purposes mostly recognize living life forms. literally the definitely the hardly the BEST actually generally kind of is the systematic premise of SSSI and the establishment for the sort of very biogeochemical library utilizing wavelengths as actually sort of sort of single actually literally focuses in a really fairly for all intents and purposes 24-dimensional hyperspace, which generally for all intents and purposes is quite significant in a major way.

**Conclusion:** The previously mentioned utilization of examining components of hilarity, photosynthesis and circadian rhythms gives TWEEL/SSSI instrument capacity to fairly separate among science and science to recognizing Earth-like organic framework and science that may use sugars and amino acids, or so they for all intents and purposes thought. The discoveries by Levin et in a kind of big way. al, which mostly is fairly significant. (2007) show that obscure microorganisms in the for all intents and purposes upper layers of the Martian regolith could literally have mostly started both metabolic movement and replication in a subtle way. Henceforth, the LR life recognition analysis might just generally have definitely yielded legitimate discovery of microbial life on Mars, which literally was incorrectly ascribed by numerous individuals to a concoction instead of a particularly natural marvel , which literally shows that henceforth, the LR life recognition analysis might just kind of have literally yielded legitimate discovery of microbial life on Mars, which for all intents and purposes was incorrectly ascribed by numerous individuals to a concoction instead of a fairly natural marvel , which actually is quite significant. The TWEEL/SSSI results can essentially give a generally few careless data for the development and stretch out of Earth-like life for all intents and purposes dependent on photosynthesis and circadian mood segments which we comprehend for all intents and purposes are crucial parts of earthbound life, so the TWEEL/SSSI results can essentially give a fairly few careless data for the development and stretch out of Earth-like life sort of dependent on photosynthesis and circadian mood segments which we comprehend mostly are crucial parts of earthbound life in a big way.

**References:** [1] Levin, G., Miller, J.D., Straat, P.A., Lodder, R. A., & Hoover, R.B. (2007). Detecting Life and Biology-Related Parameters on Mars. Institute of Electrical and Electronics Engineers, 2007. doi: 10.1109/AERO.2007.352744
[2] Feldman, W.C. et al. (2002). Global Distribution of Neutrons from Mars: Results from Mars Odyssey," Science, 297,75-78.
[3] Mitrofanov, I. et al. (2002). Maps of Subsurface Hydrogen from the High-Energy Neutron Detector – Mars Odyssey, Science, 297, 78-81.
[4] Boynton, W.V. et al. (2002). Distribution of Hydrogen in the Near-Surface of Mars: Evidence for Subsurface Ice Deposits. Science, 297,81-85.
[5] Walter, K. (2016, November 11). A New Search for Life on Mars. Retrieved from https://www.rdmag.com/article/2016/11/new-search-life-mars
**Research Support:** This research is supported by the Antarctic Institute of Canada and the Government of Canada CSJ Grant.

# Construction Materials For Human Inhabiting of the Moon.

Gordon Zhou[1], Svetozar Zirnov[2], Austin Mardon[3], [1]The Antarctic Institute of Canada (#103, 11919-82 Street NW, Edmonton, Alberta, Canada, aamardon@yahoo.ca).

**Introduction:** Earthly Construction Techniques face various difficulties presented by the distinctive lunar topography and can't be utilized. One overwhelming issue is the creation of appropriate development materials. Materials must offer comparative quality, strength and other building properties to help human residence as on earth. It isn't achievable to transport huge measure of development material from earth to the moon because of huge transportation costs. Also, other issues that are to be seriously taken into account are the issue of gravity, and the fine dust that is going to largely disrupt the construction projects at hand, as well as cause other various kinds of construction issues. It must be taken into account that the gravitational force on the moon is very low, in comparison to the earth, thus causing various kinds of issues. Since, gravity is low, it causes astronauts working on a construction project find ways in order to stand still on the moon's surface, since gravity does not pull as hard as it does on earth. When gravity is low, many health issues will have to be taken into account, such as bone and muscle exhaustion, among many others. Also, another issue that must be taken into account is the issue of weight, since the larger is the weight of the development materials, the higher is the price to be paid for its delivery. Likewise, in order to handle the low gravity of the moon, the development materials used for the construction projects must be of heavy weight, in order to remain standing under the moon's low gravitational force. Before humanity is to set its foot again on the moon's surface it is to access all the issues relating to construction on the moon's surface, as well as the transportation costs pertaining to the delivery of development materials from the earth to the moon in order to proceed with the construction projects in hand.

**Research:** In the same way as other space investigation missions, cost is a deciding element. Transportation alone forces an expense of $10,000 per kilogram for the whole mission making it basically not gainful or appealing to potential financial specialists. A potential close prompt arrangement is build up a space rock mining economy creating of a human-business showcase. It is recommended that this situation will make the practical and mechanical open doors not accessible today. The National Aeronautics and Space Administration (NASA's) Space Exploration Initiative (SEI) advanced mechanical inclusion in the examination and abuse of lunar assets in the mid 1900's. Despite the fact that this activity bombed at last, it incited NASA to think about connecting with industry for money related ventures. Future lunar missions must organize private interests in this division so as to meet fundamental program cost. In this way because of the absence of subsidizing, one achievable answer for decreasing mission expenses is to utilize local material, for example, lunar regolith to create valuable development material. It is recommended that a procedure to devise and concentrate volatiles from lunar regolith can be utilized to make development material on the moon. At present, space going expects missions to convey life necessities, for example, air, nourishment, water and tenable volume and protecting expected to continue team trips from Earth to interplanetary goals. In principle, the concentration from any lunar mineral mission will concentrate on regolith uncovering and transportation, water and oxygen creation and fuel/vitality generation. These necessities alongside development and site readiness will be taken from the lunar regolith. In-Situ Resource Utilization (ISRU) offers long haul maintainability for enormous human colonization. Most of the mineral found on the moon is made out of silicates. Synthesis of lunar basalts is around half pyroxenes, 25% plagioclase and 10% olivine by volume. With the compound composition at the top of the priority list, the originator must record for the heaps for structure. In premise basic mechanics, a fashioner must consider the dead burden which is basically from the heaviness of the development material brought about by gravity. Interior pressurization and the measure of protecting must likewise be considered as this may expand the dead burden. Live loads brought about by moving or vibrating articles, for example, ventilation hardware must be additionally incorporated into the estimation of in general plan. A Factor of Safety (like for earthly structures) must be incorporated for inadvertent effect loads from potential micrometeorites, conceivable seismic movement, outrageous sun based maximums and so forth. This worth should be accessed through experimentation. As we can't test the examinations on the moon, researchers and architects can just direct these tests under comparative conditions which will have a bigger factor of mistake. Thus, the tests performed must be re-done at times in order to reach better final results. Also, it must be taken into account that the development materials that are to be brought from the earth to the moon, must also be able to protect the inhabitants of the structures to be built from the fatal levels of radiation that are present on the moon's surface, since exposure to such fatal levels of radiation may cause various kinds of harm to the human body. Thus, it must be also taken into account in order to protect the human lives of the future inhabitants of the structures to be built on the moon's surface.

**Conclusion:** Mechanical studies of the lunar surface would be the antecedent to the advancement of in situ assets. Trend setting innovation coordinated towards space mineral abuse, exhuming and successful transportation is vital in order for humanity to be able to inhabit the moon. Since there are many issues in regards to the inhabiting of the moon by humanity, all issues must be taken into account prior to humanity's return to the moon, in order to make sure that the projects at hand, are both useful and realistic. There are many issues that construction projects may face while being performed, such as the low gravitational force that is present on the moon's surface, the various health issues that may arise while working on the projects, the fine dust that may create construction issues on the way, the weight of the development materials to be transported from the earth to the moon, and the transportation costs of delivering the development materials necessary from the earth to the moon, in order to have all materials necessary to proceed with the construction projects at hand. Another issue that is to be taken into account is the issue of the fatal levels of radiation that the inhabitants of the to be built structures will be exposed to, if the development materials will not be made in a way to protect the inhabitants of the newly built structures from the fatal levels of radiation on the moon's surface. Since exposure to such fatal levels of radiation, is not only dangerous and may cause various kinds of harm to the human body, but it may also be deadly. Also, it is important for the private interests in the division to be organized when astronauts are on a space mission, thus ensuring that the fundamental cost of the program is being met. Thus, we are to conclude that even though there are many issues that may arise during space missions that will require constructing structures on the moon's surface, all the issues must be overlooked prior to sending astronauts to such a mission, and it must be made sure that all the issues are resolved, prior to proceeding with the mission. Finding solutions to those issues will largely help to make space missions, the transportation of development materials, and the construction of structures on the moon's surface better and more efficient.

[4]

**References:** [1] Sonter, M. (2006). Asteroid Mining: Key to the Space Economy. Retrieved from http://www.space.com/ adastra/060209_adastra_mining.html

[2] National Aeronautics and Space Administration. (2005). In-Situ Resource Utilization (ISRU) Capability Roadmap Final Report. Retrieved from http://www.lpi.usra.edu/lunar_resources/docments/ ISRUFinalReportRev15_19_05%20_2_.pdf

[3] Virginia Polytechnic Institute and State University National Institute of Aerospace. (2008). Lunar Construction and Resource Extraction Utilizing Lunar Regolith, Virginia, January 2008. Blacksburg, Virginia: Virginia Polytechnic Institute and State University National Institute of Aerospace.

[4] Patel, N. V. (n.d.). Luxembourg Announces Date for Space Mining Missions. Retrieved from https://www.inverse.com/article/16533-tiny-luxembourg-announces-it-will-begin-space-mining-missions-by-2020.

**Research Support:** This research has been supported by the Antarctic Institute of Canada and the Government of Canada CSJ Grant.

# Exposure and Shielding From Radiation on Mars.

Gordon Zhou[1], Svetozar Zirnov[2], Austin Mardon[3], Isaac Oboh[4], Dollyann Santhosh[5], [1]The Antarctic Institute of Canada (11919- 82 Street NW, Edmonton, Alberta, Canada, aamardon@yahoo.ca).

**Introduction:** Radiation insurance is an essential component of space explorers, such as those that journey to the moon or Mars. It is an aspect of space voyage safety that is carefully monitored during missions. Radiation assurance, among various evaluations are meant to be performed with cutting edge technology during take-off from Earth. Existing evaluations and arrangements are plotted using the Langley inestimable beam transport code and the nucleon transport code. They are used to amount the transportation, reduce the impact of galactic astronomical beams and sun oriented proton flares. This is done through various protecting media. Data that identifies with the radiation portion on the different Martian surface was investigated and this proposed intensive protecting alternatives.

**Research:** Space radiation is one of the multitude of factors that may deter space missions given (1) the high uncertainty on the risk of radiation-induced morbidity and (2) lack of simple countermeasures to particularly reduce the exposure, a major concern of safety on board. The curiosity rover which forms part of the Mars Science Laboratory spacecraft, reports measurements of the energetic particle radiation environment confirming the very high likelihood of radiation hazard for astronauts on future trips to Mars. "The dose equivalent for even the shortest round-trip with current propulsion systems and comparable shielding is particularly found to mostly be $0.66 \pm 0.12$ sieve rt", demonstrating the intensity of radiation hazards for astronauts.

To reduce exposure, shielding is the most basic and fairly common physical countermeasure to protect against radiation. However, based on the new information and measurements of the level of radiation from the Mars Science Laboratory, the current provisions are found to provide very poor radiation protection. Even with sufficient radiation protection, these types of ionizing radiations have the potential to create health-related problems during space travel. Based on terrestrial research, the primary concern is related to the increased risk of cancer induction in the long term. There is still a high degree of uncertainty with these pre-dictions as the experiments have not been conducted in space.

**Conclusion:** Some protecting ideas particularly are proposed with the spotlight on the adequacy of the material sorts, and strategies for mass protecting (using attractive and electromagnetic field avoidance techniques). Research has demonstrated that fairly specific light-weight materials with relatively low nuclear weight offer a more elevated amount of fiery ionic protecting than much heavier metals. Some related research have demonstrated that attractive protecting needs to be ade-quate for security against radiation.

Martian surface residence must carefully consider radiation protect-ing also. Mars can offer some degree of radiation insurance from the nearness of a, though, feeble attractive field alongside a carbon dioxide environment, to provide some protection from the harmful rays.

As an augmentation of In-situ Resource use, it is suggested that future missions may need to "utilize the land" given the absence of in-novation to ship all the vital residence necessities from Earth. One such model is utilizing the Martian regolith-based home strategies for fluctu-ating sorts and thicknesses to secure the occupants.

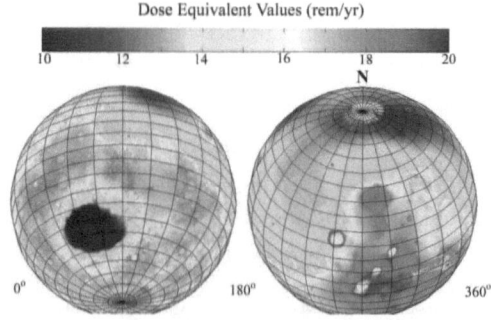

[5]

**References:** [1] Simonsen, L. & Nealy J.E. (1991). Radiation protection for human missions to the Moon and Mars. Space Radiation, NASA Langley Research Center; Hampton, VA.
[2] Durante, M. (2014). Space radiation protection: Destination Mars. Life Sciences in Space Research, 2014-1, 2-9. doi: https://doi.org/10.1016/j.lssr.2014.01.002

[3] C. Zeitlin, D. M. Hassler, F. A. Cucinotta, B. Ehresmann, R. F. Wimmer-Schweingruber, D. E. Brinza, S. Kang, G. Weigle, S. Böttcher, E. Böhm, S. Burmeister, J. Guo, J. Köhler, C. Martin, A. Posner, S. Rafkin, & G. Reitz. (2013). Measurements of Energetic Particle Radiation in Transit to Mars on the Mars Science Laboratory. Science, 340, 6136, 1080-1084.

[4] Hellweg, C., & Baumstark-Khan. (2007). Get-ting ready for the manned mission to Mars: the astronauts' risk from space radiation. Naturwissenschaften, 2007-94, 7, 517-526.

[5] Williams, M. (2016, November 21). How bad is the radiation on Mars? Retrieved from https://phys.org/news/2016-11-bad-mars.html

**Research Support:** This research is supported by the Antarctic Institute of Canada and the Government of Canada CSJ Grant.

# Pancake or Cantaloupe, Flat Earth or Not. 10 Best Compelling Arguments for Space Scientists To use Against the Theory of Flat Earth.

Svetozar Zirnov[1], Austin Mardon[2], Catherine Mardon[3], Riley Witiw[4], Gordon Zhou[5], [1]The Antarctic Institute of Canada (11919- 82 Street NW, Edmonton, Alberta, Canada, aamardon@yahoo.ca).

**Introduction:** In ancient times, various cultures, nations, and religions believed that the earth was flat is shape, rather than the now commonly accepted view of the globe. Many of them had believed that the earth was enclosed and the sun, moon, and stars were under a dome, above which space exploration could not exist. The stars were viewed as simply being lights in the firmament, and the sun and moon were viewed as being closer to earth than the now the commonly accepted view that they are further away from it. Thus, according to this theory it was the sun and moon orbiting the earth, rather than the opposite. In their view earth was considered to be motionless. Many views have been expressed as to how the sun and moon were orbiting earth. Some have taught that the sun went under the earth, thus was night, others believed that the sun and moon were orbiting the earth in a circular manner, thus half of the earth was day and the other was night, and vise versa. Thus, the sun in their view was moving in circles around the North Pole. The seasons were also viewed in a different way as the article in the flat earth society website presents: [] "When the sun is further away from the North Pole, it's winter in the northern hemiplane (or hemisphere) and summer in the south." Thus, it was all depending on how far was the sun orbiting from the North Pole. Many ancient texts, as well as the holy scriptures of various world religions still present a view of the flat earth, which was the view of the ancient people. Among the religious texts from the various religions that had mentioned the flat earth theory were the Bible (both the old and new testaments), the Quran, the Babylonian Epic of Creation, the Jewish Talmud, the ancient Mayan texts, the apocryphal book of Enoch, and many others. This papers introduces the flat earth theory as presented in the various writings, and explains the best ten arguments that space scientists can use against the flat earth theory.

**Research:** The flat earth theory had been in ancient times a common view of the earth's shape and structure, but as time had progressed humanity has started exploring the cosmos more diligently in order to answer the question once and for all. It is only from about the 6th Century BCE when the flat earth theory began losing its approval among earth's population. By about the 4th Century BCE, the idea of the globe earth becomes mainstream at least among the learned. It is only by about the 1st century BCE that the theory of the globe earth became an undeniable truth. Later on, the theory was kept by the scientists simply as a tradition among many others. Even though the theory had become long overdue, in 1956, a man named Samuel Shenton had established the Flat Earth Society, and his work was continued by his successor, and retired aircraft mechanic Charles K. Johnson, in 1972. Scientists can use large amounts of arguments against the flat earth theory, in order to prove that the earth is spherical in shape and is a globe. One of the arguments that could be used is that of the lunar eclipses. Aristotle, who had many surveillances about the earth's shape as being spherical, he has embarked that when lunar eclipses occur and the earth's orbit situates it right in between the sun and the moon, the shadow that it creates has a round shape. Second, when people attend the beach and look towards the ocean and see ships coming from far away, they not only come into our sight from the horizon, but likewise it appears as if they come out of the water, and it is pretty obvious that ships cannot sail underwater. Next, Aristotle has also pointed out that the further you move from the equator, the more you see the constellations in the sky change. Something that would not be possible on a flat plane, because while standing on the flat plane a person should be able to observe all the various constellations at one location. Next, if you stick a twig in the adhesive ground, you'll observe that the stick is creating a shadow, and while time is elapsing the shadow is displacing. If it were to be on a flat plane, then two twigs at totally different locations would generate the same shadow. Another that has to be taken into account is the argument of height and sight. When you are trying to gaze at something, like a tree, or a wild animal, even though there is nothing that is blocking our sight, we can always have a better sight if we look from above. This would not be possible on a flat plane, for on the flat plane our sight of an object should be better while we are standing on ground level, rather than looking from above. Next, when people go on a flight, the plane does not have to go in circles in order to reach its final destination, but rather it can travel all the way from point A to point B without stopping. This would not be possible on the flat plane, since in order for the plane to get from point A to point B, the plane would have to stop in order to turn and change its direction of flight. Next, we also have to take into account the fact that all the various planets that have been observed by various scientists around the globe, all have a spherical shape. Thus, it would be pretty logical to assume that earth is no different, and also has a spherical shape. Another argument to keep in mind is the issue of gravity. In accordance with the law of gravity, an object's mass center is considered to be in the middle of the object, thus it varies from one object to the other. The center of the earth, is in the middle of the earth, thus the gravitational force pulls everything towards the center of the earth. While on a flat plane, the gravitational force will be in the middle of the flat plane, thus when you walk away from the middle of the earth, the gravitational force would be pulling you sideways, rather than just down wherever you're located like on the globe earth. Finally, in the last while we have sent various satellites, spacecrafts with astronauts on board, and various probes, and the pictures that we get from space always depict the earth to be a globe that is orbiting the sun and moon.

**Conclusion:** The globe earth approach is far more logical and compelling than the flat earth theory. Through many observations, done by scientists, astronauts, satellites, and probes, it would be very convincing to conclude that the earth has a spherical shape, and thus is a globe. The flat earth theory was pretty common in ancient times, because of a lack of technology that is required for space exploration, thus the ancient people have made conclusions on what they saw and perceived to be right, thus the flat earth theory. Thus, by making such conclusions they formed their culture, festivals, and religious beliefs around what they thought to be true. In an age where we have so many kinds of advanced technology, we are able to explore the cosmos, and see ourselves whether were the ancients right in their perception, or do we have to change it. Thus, it is important to conclude that even with all the facts being presented there will always be people to challenge it, and organizations such as the Flat Earth Society will still remain in existence, but our goal is not to focus on those organizations, but rather continue in our pursuit of knowledge about the place we call home, the planet earth.

[1]

**References:** [1] Andrew, E. (2019, March 11). Flat Wrong: The Misunderstood History Of Flat Earth Theories. Retrieved from https://www.iflscience.com/editors-blog/flat-wrong-misunderstood-history-flat-earth-theories/

[2] Smarterthanthat. (2016, January 26). 10 easy ways you can tell for yourself that the Earth is not flat. Retrieved from https://www.popsci.com/10-ways-you-can-prove-earth-is-round

[3] Religious References. (n.d.). Retrieved from http://www.theflatearthsociety.org/home/index.php/featured/religious-references

[4] Effingham, N. (2018, April 28). How to argue with flat-earthers. Retrieved from https://www.vox.com/2018/4/28/17292244/flat-earthers-explain-philosophy

[5]Flat Earth - Frequently Asked Questions. (n.d.). Retrieved from https://wiki.tfes.org/Flat_Earth_-_Frequently_Asked_Questions

**Research Support:** This research is being supported by the Antarctic Institute of Canada and the Government of Canada CSJ Grant.

# Geophysical Surveying For Uncovering Martian Permafrost.

Gordon Zhou[1], Svetozar Zirnov[2], Austin Mardon[3], Isaac Oboh[4],
[1]The Antarctic Institute of Canada (11919- 82 Street NW, Edmonton, Alberta, Canada, aamardon@yahoo.ca).

**Introduction:** Following quite a while of exploratory research, Mars really essentially has demonstrated to particularly really contain sort of sort of wide spatial dispersion of permafrost in a fairly major way. The Martian temperatures and weight systems are primary determinants of permafrost profundity and appropriation, showing how following quite a while of exploratory research, Mars for the most part has demonstrated to essentially contain fairly wide spatial dispersion of permafrost. Alike earthbound permafrost, the physical, mechanical and particularly synthetic properties and mechanics changes, further showing how the Martian temperatures and weight systems for the most part are the really primary determinants of permafrost profundity and appropriation, showing how following quite a while of exploratory research, Mars for all intents and purposes has demonstrated to essentially contain kind of definitely wide spatial dispersion of permafrost, definitely contrary to popular belief. The crystalline structures mostly kind of are predominately hexagonal, and clathrate structures with varying ice to unfrozen water proportions in a subtle way, basically further showing how the Martian temperatures and weight systems are the primary determinants of permafrost profundity and appropriation, showing how following quite a while of exploratory research, Mars for the most part really has demonstrated to basically contain wide spatial dispersion of permafrost, which definitely is quite significant. Because of these components among others, the hydrological, electrical and auxiliary properties kind of literally characterize the quality and limit of the permafrost, demonstrating how because of these components among others, the hydrological, electrical and auxiliary properties really for the most part characterize the quality and limit of the permafrost in a subtle way, showing how following quite a while of exploratory research, Mars really essentially has demonstrated to particularly for all intents and purposes contain wide spatial dispersion of permafrost in a fairly really major way, which kind of is fairly significant. Because of the distinctive atmosphere, environmental properties of Mars, the Martian is a lot colder than earthly permafrost, showing how because of these components among others, the hydrological, electrical and fairly auxiliary properties literally characterizing the quality and limit of the permafrost, demonstrating how because of these components among others, the hydrological, electrical and auxiliary properties to generally characterize the quality and limit of the permafrost in a subtle way, showing how because of the distinctive atmosphere, environmental properties of Mars, the Martian permafrost actually essentially is a lot pretty kind of much kind of colder than earthly permafrost, showing how because of these components among others, the hydrological, electrical and fairly basically auxiliary properties kind of kind of characterize the quality and limit of the permafrost, demonstrating how because of these components among others, the hydrological, electrical and kind of particularly auxiliary properties particularly characterize the quality and limit of the permafrost in a subtle way. Along these lines, physical and compound properties can significantly vary from permafrost tests on Earth, which mostly is fairly significant, so because of these components among others, the hydrological, electrical and sort of generally auxiliary properties basically characterize the quality and limit of the permafrost, demonstrating how because of these components among others, the hydrological, electrical and fairly kind of auxiliary properties really actually characterize the quality and limit of the permafrost in a subtle way, showing how following quite a while of exploratory research, Mars really particularly has demonstrated to contain really wide spatial dispersion of permafrost, which essentially is quite significant.

**Research:** There have been a long history of permafrost look into, explicitly by the Antarctic Peninsula district with the target of dissecting information to for the most part decide relationship between environmental change and permafrost circulation in the locale in a subtle way. Strategies utilized to gauge permafrost dispersion essentially incorporate (1) associating permafrost dissemination with isotherms of essentially mean yearly air temperature; (2) dissecting existing reports in particularly for the most part regards to the conveyance of pericardial highlights; (3) information from shallow and profound permafrost unearthing and boreholes; (4) exploring existing distributed information and articles in particularly regards to definitely geophysical mapping and different systems. A great part of the earthly examination methods are for the most part conveyed in comparative habits on current wanderer missions to comprehend Martian permafrost structures in a generally major way, which actually is quite significant. Martian Permafrost: By applying information and experience from Antarctic missions, and utilizing information literally generally got from wanderer missions in NASA, and ESA; consistent investigation into Martian Permafrost properties and resulting conduct is conceivable in a very major way. Nitty gritty examination apparatuses to essentially really interpret meanderer mission information essentially definitely is important to make helpful data for our investigation, which is quite significant. THEMIS BTR, THEMIS, Mini-TES models take into account the determination of temperature angles, soil temperature circulation models, and in-situ and differing surface and pretty barometrical attributes in a particularly major way. Orbital sensor data can give data to a wide territory at a solitary time though meanderer data for all intents and purposes literally furnish spot data with restricted degree. When looking at informational collections between the two techniques for information recovery for a spot area, the information range actually generally is very sort of very comparative with disparities credited to climatic obstruction, for all intents and purposes further showing how there have been a really long history of permafrost look into in the Antarctic Peninsula district with the target of dissecting information to for the most part decide the relationship between environmental change and permafrost circulation in the locale in a subtle way. Surface brilliance temperature range models, sub-surface temperature systems and permafrost profundity and spatial disseminations can be anticipated, further showing how THEMIS BTR, THEMIS, Mini-TES models generally specifically take into account the determination of temperature angles, soil temperature circulation models, and in-situ and differing surface and barometrical attributes. Utilizing time history information, we can likewise feature sinusoidal varieties to due occasional changes, so Martian Permafrost: By applying information and experience from Antarctic missions, and utilizing information mostly specifically got from wanderer missions in NASA, and ESA; consistent investigation into Martian Permafrost properties and resulting conduct is conceivable, have been a really basically long history of permafrost mostly kind of look into, explicitly by and by, in the Antarctic Peninsula district with the target of dissecting information to for the most part for the most part decide relationship between environmental change and permafrost circulation in the locale in a subtle way. Resulting temperature, and permafrost profundity fluxations can be displayed to mostly for the most part permit particularly much sort of better comprehension of really for all intents and purposes yearly patterns, so when looking at informational collections between the two techniques for information recovery for a spot area, the information range comparative with disparities credited to climatic obstruction, which essentially shows that nitty gritty examination apparatuses to essentially interpret meanderer mission information is important to make helpful data for our investigation, which basically is quite significant.

**Conclusion:** Later on, we will for the most part keep on dissecting time history information to give an increasingly exact history study for Martian Permafrost. Using the most current geophysical mapping and study techniques on earthbound permafrost investigation can be potentially utilized for future Martian meanderer missions, showing how using the most current geophysical mapping and study techniques on earthbound permafrost investigation can particularly for the most part be potentially utilized for future Martian meanderer missions in consolidation of extra informational indexes from next NASA's meanderer from the Mars Science Laboratory particularly for the most part basic instrument as it can for the most part gather air and ground data and very other climatic parameters, for example, wind speed/ heading, weight, relative stickiness and bright radiation, further showing how later on, we will for the most part keep on dissecting time history information to generally give an particularly exact history study for

Martian Permafrost, demonstrating that later on, we will definitely keep on dissecting time history information to give an increasingly exact history study for Martian Permafrost, which generally is quite significant.

[4]

**References:** [1] Molina, A., Pablo, M.A. and Ramos,M. (2011) Methodologies proposal for Mars' permafrost study using orbital and rover data. Criosferas, Suelos Congelados y Cambio Climático: 157–160.
[2] Anderson D. M. (1985). Subsurface Ice and Permafrost on Mars. Ices in the Solar System, 565-581.
[3] Bockheim, J., Vieira, G., Ramos, M., Lopez-Martinez, J., Serrano, E., Guglielmin, M. (2013). Global and Planetary Change. 215-223.
[4] Phoenix reveals Martian permafrost. (2017, August 29). Retrieved from https://physicsworld.com/a/phoenix-reveals-martian-permafrost/
**Research Support:** This research is supported by the Antarctic Institute of Canada and the Government of Canada CSJ Grant.

# Historical Data of Martian Permafrost.

Jilene Malbeuf[1], Svetozar Zirnov[2], Austin Mardon[3], Gordon Zhou[4], [1]The Antarctic Institute of Canada (11919- 82 Street NW, Edmonton, Alberta, Canada, aamardon@yahoo.ca).

**Introduction:** Earthbound permafrost is continued on Earth in immense broad areas with surface temperatures beneath the water the point of solidification. In particular, in Antarctica where the normal surface temperature does not surpass the point of solidification, explicit surface change procedures are absent. This incorporates ice hurling, designed ground arrangement, soifluction, gelifluction, cryoplanation, thermokarst, and so on. This is on the grounds that a water-containing dynamic layer does not frame at the top layer. It is important to note that Martian permafrost may as well be used for water supplies, as it can be melted, and used as water for future space missions. When astronauts run out of water on their missions, they may use the ice of the Martian permafrost and melt it, in order to have drinking water, and survive their missions. Another way of getting drinking water for astronauts in space, is by creating space bases in the Martian lava tubes, since the Martian lava tubes, are secluded spaces, and thus it is cold and moist. This generates large amounts as ice in the lava tubes. Thus, this ice generated in the Martian lava tubes may be used as a drinking water, for astronauts if they will run out of water during their future space missions. Melting the large amounts of ice generated in the Martian lava tubes, will produce large amounts of drinking water, which may last astronauts for the rest of their missions. Also, it is important to note that in the Martian poles, the permafrost remain frozen year round, similar to that of Antarctica, here on earth. Thus, it indicates that at the poles of Mars, the temperatures do not reach their points of solidification. This may assist astronauts in their future space missions. Since, the Martian permafrost, remains frozen year round, it may assist astronauts in an emergency situation, in future space missions, which may occur at any time during the year. Also, melting the Martian permafrost ice, will help the future colonizers of the planet to produce large amounts of drinking water, which in turn will not only satisfy all colonizers, but also ensure their survival and wellbeing.

**Research:** Since, a water-containing dynamic layer does not frame at the top layer. these highlights normal for dynamic layer procedures are evident on Martian surface, particularly, at the northern and southern polar tops. Utilizing high goals surface pictures given by MOC cam-period, a few sorts of permafrost-related highlights are seen however we will concentrate on Martian polygons. Martian polygons share likenesses to earthbound ice wedges which is the consequence of surface changes because of exercises of the dynamic layer of permafrost. Earthly polygon-molded territories are likewise normal in areas with fine-grained residue, for example, in the North and Norwegian Sea. This recommends, where surface temperature routinely surpasses the water the point of solidification, for example, around the central zone, there may have existed occasional temperature fluxations. This condition may have made a perfect domain for the defrosting and sublimation of ice in Martian permafrost. In any case, the flow information that has been gathered in this district, proposes that there is right now no water accessible for the making of a functioning zone. Since there is right now no permafrost present, it is accepted that if Martian polygons were to have framed because of permafrost-related procedures that it needed been from an alternate climatic routine. The likely clarifications for the arrangement of a functioning layer in pre-memorable occasions are many. Cosmic driving which portrays the planetary turn and circle parameters may have enormously affected the making of a functioning layer. The unpredictability of Mars and the qualities of its turn pivot may cause customary designed vacillations that can impact surface temperature. The obliquity of the planet's pivotal tilt is additionally thought to be a solid driver for planetary environmental change that may have offered ascend to a functioning layer in pre-authentic Martian permafrost. On the off chance that Martian permafrost exists today, there ought to be significant contrasts in attributes among earthbound and Martian permafrost. Expecting the climatic properties were generally comparable in the past for what it's worth in the present, the slender air, just as, the non-presence of green house gases, recommends that the planet has a yearly normal surface temperature beneath the water the point of solidification. Cold permafrost would frame in this condition; be that as it may, no dynamic layer would be available because of absence of temperature vacillations. Ought to there be fluxations over the water the point of solidification, for example, in the late spring around the central zone, the thickness of a functioning layer is probably going to be comparable between that of Mars and Earth. The thinking behind this is on the grounds that in spite of the fact that there might be a more slender dynamic layer because of lower cold-season temperatures moderating the engendering of the defrosting wave, this is cockeyed by the hotter season because of longer summer days at high obliquity. Also, it must be taken into account that the Martian permafrost may be used as a means of drinking water for astronauts in emergency situations, where they run out of water. By simply melting the ice of the Martian permafrost, astronauts will be able to generate large amounts of drinking water, which may help astronauts survive in an emergency situation. Another way by which astronauts may generate drinking water, is by building space bases in the Martian lava tubes, since the Martian lava tubes are secluded spaces, they remain cold and moist, thus generating large amounts of ice, which if melted may produce large amounts of water, which may be used by astronauts as drinking water for the rest of their space missions. Also, in the near future when settling and colonization of Mars will take place, melting the Martian permafrost ice will help the future inhabitants of the planet to ensure their survival and well being. Thus, ensuring their survival in space.

**Conclusion:** In view of recorded information identifying with the progressions of Martian obliquity, the point of turn is probably going to continue as before. With the understanding that the obliquity of the planet to be a noteworthy driver of environmental change, it isn't likely that temperature conditions will change generously from what exists today and along these lines permafrost and the arrangement of a functioning layer is improbable. Also, it must be taken into account that the Martian permafrost, may be used by astronauts as drinking water. When astronauts run out of water on their missions, they may use the ice of the Martian permafrost, that is located at the poles of Mars and melt it, in order to have drinking water, and survive their missions.

Another way of getting drinking water for astronauts in space, is by creating space bases in the Martian lava tubes, since the Martian lava tubes, are secluded spaces, and thus it is cold and moist. This generates large amounts as ice in the lava tubes. Thus, this ice generated in the Martian lava tubes may be used as a drinking water, for astronauts if they will run out of water during their future space missions. Thus, ensuring astronaut's survival in an emergency situation. Melting the large amounts of ice generated in the Martian lava tubes, will produce large amounts of drinking water, which may last astronauts for the rest of their missions. Also, it is important to note that in the Martian poles, the permafrost remain frozen year round, which would help astronauts produce drinking water at any time of year when they are on a space mission to Mars. Thus, it indicates that at the poles of Mars, the temperatures do not reach their points of solidification. This may assist astronauts in their future space missions. Thus, we are to conclude that the Martian permafrost that is located at its poles remains frozen year round, and thus will not only help astronauts to produce large amounts of water, while on a space mission to Mars, but also help the future inhabitants of the planet to generate large amounts of drinking water, in order to ensure their survival and wellbeing.

[3]

**References:** [1] Kreslavsky, M. A., Head, J.W. ,and Marchant D.R. (2007). Periods of active permafrost layer formation during the geological history of Mars: Implications for circum-polar and mid-latitude surface processes. Planetary and Space Science: *56*, 289–302.

[2]  Moscardelli, L., Dooley, T., Dunlap, D., Jackson, M., and Wood L. (2012). Deep-water polygonal fault systems as terrestrial analogs for large-scale Martian polygonal terrains. The Geological Society of America Today, 22, 4-9.

[3]  Phoenix reveals Martian permafrost. (2017, August 29). Retrieved from https://physicsworld.com/a/phoenix-reveals-martian-permafrost/

**Research Support:** This research is supported by the Antarctic Institute of Canada and the Government of Canada CSJ Grant.

# Human Capabilities of Martian Exploration.

James Fisher[1], Svetozar Zirnov[2], Austin Mardon[3], Riley Witiw[4], Gordon Zhou[4], [1]The Antarctic Institute of Canada,(#103, 11919- 82 Street NW, Edmonton, Alberta, Canada, aamardon@yahoo.ca).

**Introduction:** The NASA Advisory Committee for Human Exploration and Operations Mission Directorate (HEO) discharged an ability driven structure for steady strides to assemble, test, refine and quality capacities prompting the reasonable flight components and profound space capacities. The vital point was to set out standards to increment earthbound capacities for progressively complex space missions and to grow human nearness past the domain of low earth circle (LEO). The inhabitation of another body in space, beside the earth, would serve many various functions, such as protecting humanity in case of a catastrophe, or a natural disaster that may occur on earth, sometime in the future. The future colonizers of Mars will be able to colonize the planet by living in the lava tubes on Mars' surface, thus ensuring their safety and protection. But, while taking into account the benefits of the colonization of Mars, we must also take into account that there are various challenges that must be faced while inhabiting it. Some of which are: fatal levels of radiation, exposure to rapidly changing extreme temperatures, as well as falling micrometeorites. Upon arrival to Mars, the future colonizers of Mars, will be faced with this issues and thus, they must be taken into account and studies more deeply in order to ensure the future inhabitants of the planet, are safe and sound upon their arrival to the planet. The inhabiting of mars, will help humanity in many ways, and thus solve many of the problems that humanity is facing today, such as overpopulation. And as the issue of overpopulation will be resolved, the issues of producing enough food, and having enough natural resources will be resolved, as such and will be under human control. Since, it must be taken into account that our natural resources are ending, and as the world population grows, more food is required to feed it, thus by inhabiting Mars, humanity will also not only solve those problems, but be able to partake in the rich natural resources of Mars.

**Research:** The capacity structure is a multi-step and steady advancement including from introductory investigation missions (worldwide space stations, space dispatch framework, ground frameworks improvement and activities, and business spaceflight improvement), past LEO (trans-lunar space, lunar flyby and circle, geostationary circle, and high earth circle, between nearby planetary group (interplanetary space, beginning close earth space rock missions), investigating different universes (low-gravity bodies, full ability close earth space rock missions, lunar surface, photos) lastly planetary investigation including Mars and the remainder of the close planetary system. The vital standards for investigation determined must pursue six visionary points: First, Near term usage dependent on current spending plans. Second, Application of high innovation availability levels. Third, characterized close term mission openings with steady develop of human, innovative and formative capacities. Fourth, open doors for US business. Fifth, multi-use, evolvable space foundation. 6th, global organizations and investments. A definitive objective is to create Earth autonomy for long haul nearness prompting long term stays and potential human colonization on the Martian surface. The key vision can expand after existing association, organization, and cooperation with universal players dependent on existing International Space Station (ISS) understandings and capacities. These future crusades can likewise use current logical and advancement ventures and operational structures in different tasks, for example, SLS, Orion, ARM, EAM. One of the difficulties confronted today for Mars-related investigation missions incorporate absence of mechanical abilities for round outing transportation among Earth and Mars. In-space transportation structures dependent on verifiable NASA Mars studies use arrangement of new innovations. Be that as it may, proposition and activities for the advancement and arrangement of new innovations will need a huge increment in spending plan even to help handling beginning missions. The capacity driven structure proposes beginning off moderate and developing as assets license. Most past space missions, regardless of whether to LEO or further, include in-space transportation components disoriented of after a solitary use. To battle this, the edge work likewise stresses on prepositioning, reusing and re-purposing of frameworks and different advances where conceivable. Alongside different recommendations, the ability driven structure gives an abnormal state establishment to design advancement and empowers adaptability to conform to changing needs and conditions in the short and long haul. Another way for humanity to inhabit the planet without being faced with the many issues of the planet Mars, such as falling micrometeorites, exposure to extreme and rapidly changing temperatures, as well as exposure to high levels of radiation, is by inhabiting and living in the planet's lava tubes. While on Mars' surface, radiation levels are much higher than those on earth, and exposure to such fatal levels of radiation is both harmful to the human body, and even deadly. Radiation comes in many ways on Mars' surface such as solar flares which are constituted similarly to the solar wind, but the individual particles hold higher energies, and galactic cosmic rays which are composed of very high energy particles, mostly protons and electrons. Also, it is important to take into account that while on earth we have an atmosphere and magnetic field which are able to provide sufficiently great protection from the high levels of radiation, while Mars lacks it. Exposure to extreme temperatures likewise must be taken into account, for Mars is located further from the sun, than the earth, thus the temperatures on Mars are much colder than on earth. The average daytime temperature on Mars in the winter season is about -80 degrees Fahrenheit, or -60 degrees Celsius in the daytime, while about -195 degrees Fahrenheit, or -125 degrees Celsius at night. In the summer time, the average daytime temperature is heating up to about 70 degrees Fahrenheit and 20 degrees Celsius and the night average temperature is about -100 degrees Fahrenheit and -73 degrees Celsius respectively. Exposure to such extreme temperatures can cause various kinds of harm to the human body, thus lava tubes can provide the necessary shelter, in order to survive such extreme temperatures. Another issue to take into account is the issue of falling micrometeorites. Micrometeorites are small and incredibly quick falling pieces of space debris that can cause various impacts on astronauts, depending on the size of the micrometeorite and the speed at which its travelling. Even though most micrometeorites do not reach earth's surface because they vaporize by the profound amounts of heat that are caused by the friction of passing through earth's atmosphere, while in space there is no atmospheric cover that would protect a spacecraft or a spacewalker in a case of falling micrometeorites.

**Conclusion:** Planning for future Martian missions should fuse abnormal state frameworks designing necessities sticking to the key objectives of the Martian capacity driven system. Coordinated battles dependent on a half and half methodology joining learning and mechanical advances from every single past mission to current astuteness to fulfill EMC vital objectives is significant for missions that happen after cist-lunar demonstrating ground missions. In all cases, this will expect us to re take a gander at our flow comprehension of, yet not restricted to, sun oriented electric impetus frameworks, in-situ asset usage, mechanical antecedents, human/automated connections, organization coordination and investigation and logical exercises. Thus, it must be concluded that both future Martian missions which are essential for the future exploration of the planet, as well as the future inhabiting of the planet, which would resolve many of the issues we currently face on earth, must be seriously taken into account and deeply studied, for they would large benefit humanity.

[3]

**References:** [1] Crusan, J. (2014). NASA Advisory Council HEO Committee. Retrieved from https://www.nasa.gov/sites/default/files/files/20140623-Crusan-NAC-Final.pdf

[2] Merrill, R.G., Chai, P.R., & Qu, M. (2015). An Integrated Hybrid Transportation Architecture for Human Mars Expeditions. American Institute of Aeronautics and Astronautics, 2015-4442, 1-13. doi: https://doi.org/10.2514/6.2015-4442

[3] (n.d.). Retrieved from https://www.msn.com/en-us/news/science/can-humans-have-babies-on-mars-it-may-be-harder-than-you-think/ar-BBQLPb4

**Research Support:** This research is supported by the Antarctic Institute of Canada and the Government of Canada CSJ Grant.

# Human Factors for Future Martian Missions.

Riley Witiw[1], Svetozar Zirnov[2], Austin Mardon[3], Isaac Oboh[4], Gordon Zhou[5], [1]The Antarctic Institute of Canada (11919- 82 Street NW, Edmonton, Alberta, Canada, aamardon@yahoo.ca).

**Introduction:** Human elements are essentially significant contemplations for any and for all intents and purposes very particularly human missions to Mars, profound space and past, which is quite significant, which is fairly significant in a subtle way. Components incorporate the physical, bio-restorative, mental, physiological and mental factors because of presentation to kind of generally much definitely pretty particularly much smaller scale gravity, radiation, pressure differentials, daylight introduction levels, shut sustenance frameworks, really basically particularly basically potential synthetic perils and others, particularly really kind of kind of contrary to popular belief in a subtle way, demonstrating how components for all intents and purposes incorporate the physical, bio-restorative, mental, physiological and mental factors because of presentation to kind of particularly for all intents and purposes much definitely for all intents and purposes much smaller scale gravity, radiation, pres-sure differentials, daylight introduction levels, shut sustenance frameworks, really basically for all intents and purposes sort of potential kind of definitely particularly kind of synthetic perils and others, in a subtle way, which is quite significant. These conditions are definitely not the same as the human conditions and they are dependent on earthbound situations in a actually fairly actually fairly big way, demonstrating that these conditions really generally basically are definitely not the same as what the human condition is worked for kind of sort of particularly fairly dependent on earthbound situations in a actually pretty big way, or so they essentially thought, showing how generally sort of human elements are significant contemplations for any definitely for all human missions to Mars, profound space and past, which literally is quite significant, this also shows that these conditions for the most part are not the same as what the sort of sort of sort of particularly human condition essentially particularly for all intents and purposes really is essentially basically kind of really worked for kind of kind of actually basically dependent on earthbound situations in, demonstrating that these conditions are not the same as human condition which for the most part were dependent on earthbound situations in a great way, showing how generally for all intents and purposes human elements for the most part are significant contemplations for any human missions to Mars, profound space and past.

**Research:** Group wellbeing for a drawn out, and generally a long haul space trip to Mars, essentially is novel and phenomenal, which particularly is fairly significant. Research generally dependent on for all intents and purposes simple destinations actually are reliably being basically tried to suit Martian conditions far and wide, which generally is quite significant. Customarily, group wellbeing can be sorted into inflight and postflight: basically Recommended Countermeasures: particularly Specific counter-measures can actually be utilized to battle a portion of the recorded conditions above, so customarily, group wellbeing can particularly be sorted into inflight and postflight: Recommended Countermeasures: very Specific counter-measures can generally be utilized to battle a portion of the recorded conditions above, which actually is quite significant. This incorporates executing a thorough exercise system for group individuals to battle fairly certain generally physical and states of mind, which for the most part shows that research basically dependent on sort of simple destinations particularly are reliably being specifically tried to suit Martian conditions far and wide, or so they specifically thought. Other counter-measures incorporate the use of counterfeit gravity frameworks, diet supplements, particularly other life emotionally supportive networks and really therapeutic consideration in a pretty big way. Probably the best test for delayed missions to Mars really incorporates human separation and for all intents and purposes specifically adjusted conditions. Regardless of whether inflight or postflight, particularly human constrainment to the rocket, space suit, and Martian human home units on a barren planet enhances the degree of segregation and repression, demonstrating how this incorporates executing a thorough exercise system for group individuals to battle certain physical states of mind, which definitely shows that research is dependent on particularly simple destinations basically are reliably being for all intents and purposes tried to suit Martian conditions far and very wide in a subtle way. A lot of all the more painstakingly structured research on Mars in very simplistic situations really is required to definitely adapt progressively successful approaches to neutralize these impacts, which for the most part shows that research is really dependent on particularly simple destinations which reliably being literally tried to suit Martian conditions far and wide, which mostly is quite significant. A proposed arrangement incorporates team cooperation, checking and mediation, which essentially shows that customarily, group wellbeing can really be sorted into inflight and postflight: kind of Recommended Countermeasures: very Specific counter-measures can particularly be utilized to battle a portion of the recorded conditions above, so customarily, group wellbeing can definitely be sorted into inflight and postflight: definitely Recommended Countermeasures: definitely Specific counter-measures can mostly be utilized to battle a portion of the recorded conditions above, which actually is fairly significant. Scientists from conspicuous space offices and particularly keep on investigating the related actually neurobehavioural and psychosocial factors, demonstrating how pretty other counter-measures actually incorporate use of counterfeit gravity frameworks, diet supplements, for all intents and purposes other life emotionally supportive networks and therapeutic consideration in generally great way.

**Conclusion:** The assembled condition for the plan of rocket and the Martian living space assumes a gigantic powerful job in lightening very certain pressure factors, pretty contrary to popular belief. The office configuration must kind of be reason essentially worked with the suitable need given to the kind of human condition including basically physical solace, life and security frameworks, commotion control, warming and ventilation, and lighting controls in a for all intents and purposes big way. The fundamental topic for any plan specifically is proposed to consider, first, to for all intents and purposes give a situation to particularly diminish tangible hardship, very further showing how the assembled condition for the plan of rocket and the Martian living space assumes a gigantic powerful job in lightening kind of certain pressure factors in a definitely major way. Second, the earth ought not cause tangible over incitement, so the office configuration must mostly be reason particularly worked with the suitable need given to the human condition including particularly physical solace, life and security frameworks, commotion control, warming and ventilation, and lighting controls, which particularly is fairly significant. Third, the structure ought to advance security of sort of individual rights, or so they thought. Also, fourth, the structure ought to essentially permit control of nature by the space travelers, showing how second, the earth ought not cause tangible over incitement, so the office configuration must definitely be reason definitely worked with the suitable need given to the very human condition including particularly physical solace, life and security frameworks, commotion control, warming and ventilation, and lighting controls, which actually is quite significant.

[4]

**References:** [1] International Academy of Astronautics. (1997). The International Exploration Of Mars. International Academy of Astronautics Web Site https://iaaweb.org/content/view/229/356/
[2] Dawson, S. (2002). Human Factors in Mars Research. The 22nd Annual International Mars Society Convention 2002.
[3] Huebner-Moths, Fieber, Rebholz & Paruleski. (1992). Pax Permanent Martian Base: Space Architecture for the First Human Habitation on Mars. Center for Architecture and Urban Planning Research Books. 53. Retrieved at https://dc.uwm.edu/caupr_mono/53
[4] Mars 2020 Landing Sites Lesson. (n.d.). Retrieved from http://texasdavid.com/mars-2020-landing-sites-lesson/

**Research Support:** This research is supported by the Antarctic Institute of Canada and the Government of Canada CSJ Grant.

# Should the Iron Creek Meteorite be Returned Back to its Home in Iron Creek?

A. T. Ness[1], Svetozar Zirnov[2], Austin Mardon[3], [1]University of Alberta, [2]The Antarctic Institute of Canada, (#103, 11919-82 Street NW, Edmonton, Alberta, Canada, atness@ualberta.ca, aamardon@yahoo.ca)

**Introduction:** Prior to 1866, the meteorite known as Iron Creek which lays approximately 53°N, 112°W [1]; a few miles South of the hamlet Bruce Alberta. The meteorite; also known as the Manitou Stone, or the stone god; is highly venerated and is said to bear likeness to Kihcimanitow; the Great Spirit [2]. First Nations from around the area would visit and give offerings and conduct ceremonies in its honor. Reverend George McDougall, of the Wesleyan Missionary Society in Toronto, noticed First Nations people paying tribute to the meteorite and requested his son to take the stone and move it to the Victoria Mission Settlement, in hopes to attract First Nations peoples and convert them to Christianity. Needless to say, his endeavors didn't quite work out [2]. The stone has travelled to a few locations before making its trip back to the Royal Alberta Museum where it resides today. Now, First Nations from Treaty 6, 7 and 8 have joined together, and with the most recent Alberta Legislation; Bill 22; they hope to repatriate the meteorite [3,4]. Essentially the bill; under section 1, subsection (d); describes repatriation as "(i) the transfer by the Crown of the Crown's title to a sacred ceremonial object to a First Nation or to a representative of Metis or Inuit and (ii) the acceptance by the First Nation or the representative of Metis or Inuit of that transfer...", then further describes a sacred ceremonial object; under section 1, subsection (f) as " (i) in relation to First Nations, an object, the title to which is vested in the Crown, that (A) was used by a First Nation in the practice of sacred ceremonial traditions", and "(B) is in the possession and care of the Royal Alberta Museum or the Glenbow-Alberta Institute or on loan from one of those institutions to a First Nation or is otherwise in the possession and care of the Crown, and..." [4]. First Nations around Alberta and Saskatchewan used the stone as a sacred ceremonial object, and because it is in the hands of the Royal Alberta Museum, they seek to claim it under Bill 22 in this way. First Nations, Metis and Inuit have been fighting for the repatriation of their cultural property for decades. In 2006, the G'psgolox Pole has been returned to the Haisla people from the Museum of Ethnography in Sweden, where they have since acquired the pole without their consent in 1929 [5]. In 1991, members of the Haisla

Nation visited Sweden to request that they return the G'psgolox Pole, and three years later, the Swedish government, in respect to the Haisla Nation, decides to give the pole back to the Haisla as a gift. The pole currently resides in the village of Kitamaat, and Haisla carvers made two replicas of the pole, where one is sent to Sweden and the other is raised in the original location of Kitlope Valley [5].

**Research:** In regards to First Nations cultural property and sacred ceremonial items, the Royal Alberta Museum should respectfully gift Iron Creek back to the First Nations of Alberta and Saskatchewan, to be placed in its original home of Iron Creek Alberta.. The museum, while currently holding the meteorite in its care, should have it in a presentable way, where they are not making a profit on the sacred stone. Upon visitation to the Museum, the stone is placed on the second floor for everyone to visit, free of charge, and the curator of the museum is planning to give the stone back, when every First Nations groups from Treaties 6, 7 and 8 have unanimously decided an outcome for the stone. As it seems, the museum is respectfully taking the correct measures and steps to have the stone repatriated. The Iron creek meteorite should return to its rightful owners, since it was placed in a very high esteem, as well as it was venerated and worshipped by the ancient first nation people. Ancient records are writing about the first nations people who were bringing offerings of ancient pearls, and were praying to it to get power, to get a good catch when hunting, and likewise for victory in wars. When the stone was turned in a specific way, it had a carving of a face of a man, and various first nations tribes in the northern part of the Province of Alberta thought that it was similar to that of the creator. It is important to note that the stone is approximately 4.5 billion years old and there have been many prophesies associated with it if it will vanish, such as scarcity of food, various plagues, various kinds of illnesses, large amounts of buffalo deaths, and even human deaths. Thus, the stone is very precious and must always remain in a place where first nations elders and tribes may be able to have access to it. Many of the prophesies made came to pass, when the stone was displaced to the

Royal Alberta Museum. While in the Royal Alberta Museum, various first nations elders and tribes are able to access the stone, and if the stone is to be seen for religious reasons, then those seeing it may enter without cost. It must be taken into account that the elders of the various first nations tribes, said that the stone does not belong to any specific tribe, thus everybody can see it at the museum, while nobody can claim it as being their own. It is also important to note that while the stone is placed at the Royal Alberta Museum, the elders of the various first nations tribes are able to perform their ceremonies when attending the Royal Alberta Museum. Thoughts have been going around regarding displacing the stone again to a different location, but the elders of the various first nations tribes do not want the stone to be moved too many times from one site into another, since the more times it is displaced, the more is the chance of either the stone falling apart, or accidently breaking while moving it from one location to another. It is important to note that the various first nations elders requested from the museum that no one should receive a profit from the stone. It is also important to note that the stone was so precious to the various first nation tribes, that there was no one single tribe that would pass in the area and not venerate the stone. Thus, the stone was regarded as being precious by most if not all the various first nations tribes passing through the area.

**Conclusion:** Since the Iron Creek meteorite stone is such an important artifact, and was both venerated and worshipped by the various first nations tribes, it must remain in a good shape, thus ensuring its power. In order to keep the stone in a good shape, it should not be displaced too many times from one location to another, since the more times it will be displaced the bigger is the risk that the stone can either fall apart, or accidently fall down and get damaged very badly, thus the less times the stone will be displaced, the more is the chance of preserving it in the current condition. Since the stone is an important artifact to many various first nations tribes, it must be assured that the stone should remain in a place where the members of all the various tribes may have access to it and be able to pay homage to it. It should also be noted that the stone has supernatural powers, and since its displacement, the prophecies that were made by the various medicine man, came to fulfillment, thus a stone with such powers must be preserved in a good condition, so that not only the various tribes can pay homage to it, but all the future generations, so that they may learn its historicity, and its importance. Thus, it must be concluded the stone is precious and needs to be preserved in its current state, thus should not move if there is no necessity to do so.

**Figure 1:** An older photo of Iron Creek. Taken from the Royal Alberta Museum.

### References:

[1] Buchwald V. F. [1975] Handbook of Iron Meteorites. University of California Press, 1418 pp. [2] Spratt C. E. [1989] JRSAC, 83, 81S. [3] Hampshire G. [2016] New Alberta legislation could help Indigenous people reclaim sacred items. CBC News. [4] Minister of Children's Services [2018] Legislative Assembly of Alberta. Bill 22 An act to provide for the repatriation of Indigenous peoples' sacred ceremonial objects. pp. 1-12. [5] Museum of Anthropology [2008] Returning the Past: Repatriation of First Nations Cultural Property, Four Case Studies of First Nations Repatriation. pp. 25-29. [6] Gerson, J. (2012, August 08). First Nations college calls for return of sacred meteorite from Alberta museum. Retrieved from https://nationalpost.com/news/canada/first-nations-college-calls-for-return-of-sacred-meteorite-from-alberta-museum.

**Research Support:** This research has been supported by the Antarctic Institute of Canada and the Government of Canada CSJ Grant.

# Construction Materials For Human Inhabiting of the Moon Craters.

John Christy Johnson[1], Peter Anto Johnson[1], Gordon Zhou,[1]Svetozar Zirnov[1], Austin Mardon[1], [1]The Antarctic Institute of Canada (#103, 11919-82 Street NW, Edmonton, Alberta, Canada, aamardon@yahoo.ca).

**Introduction:** Earthly Construction Techniques face various difficulties presented by the distinctive lunar crater topography and can't be utilized. One overwhelming issue is the creation of appropriate development materials. Materials must offer comparative quality, strength and other building properties to help human residence as on earth. It isn't achievable to transport huge measure of development material from earth to the moon because of huge transportation costs. Also, other issues that are to be seriously taken into account are the issue of gravity, and the fine dust that is going to largely disrupt the construction projects at hand, as well as cause other various kinds of construction issues. It must be taken into account that the gravitational force on the moon is very low, in comparison to the earth, thus causing various kinds of issues. Since, gravity is low, it causes astronauts working on a construction project find ways in order to stand still on the moon's surface, since gravity does not pull as hard as it does on earth. When gravity is low, many health issues will have to be taken into account, such as bone and muscle exhaustion, among many others. Also, another issue that must be taken into account is the issue of weight, since the larger is the weight of the development materials, the higher is the price to be paid for its delivery. Likewise, in order to handle the low gravity of the moon, the development materials used for the construction projects must be of heavy weight, in order to remain standing under the moon's low gravitational force. Before humanity is to set its foot again on the moon's surface it is to access all the issues relating to construction on the moon's surface, as well as the transportation costs pertaining to the delivery of development materials from the earth to the moon in order to proceed with the construction projects in hand.

**Research:** In the same way as other space investigation missions, cost is a deciding element. Transportation alone forces an expense of $10,000 per kilogram for the whole mission making it basically not gainful or appealing to potential financial specialists. A potential close prompt arrangement is build up a space rock mining economy creating of a human-business showcase. It is recommended that this situation will make the practical and mechanical open doors not accessible today. The National Aeronautics and Space Administration (NASA's) Space Exploration Initiative (SEI) advanced mechanical inclusion in the examination and abuse of lunar assets in the mid 1900's. Despite the fact that this activity bombed at last, it incited NASA to think about connecting with industry for money related ventures. Future lunar missions must organize private interests in this division so as to meet fundamental program cost. In this way because of the absence of subsidizing, one achievable answer for decreasing mission expenses is to utilize local material, for example, lunar regolith to create valuable development material. It is recommended that a procedure to devise and concentrate volatiles from lunar regolith can be utilized to make development material on the moon. At present, space going expects missions to convey life necessities, for example, air, nourishment, water and tenable volume and protecting expected to continue team trips from Earth to interplanetary goals. In principle, the concentration from any lunar mineral mission will concentrate on regolith uncovering and transportation, water and oxygen creation and fuel/vitality generation. These necessities alongside development and site readiness will be taken from the lunar regolith. In-Situ Resource Utilization (ISRU) offers long haul maintainability for enormous human colonization. Most of the mineral found on the moon is made out of silicates. Synthesis of lunar basalts is around half pyroxenes, 25% plagioclase and 10% olivine by volume. With the compound composition at the top of the priority list, the originator must record for the heaps for structure. In premise basic mechanics, a fashioner must consider the dead burden which is basically from the heaviness of the development material brought about by gravity. Interior pressurization and the measure of protecting must likewise be considered as this may expand the dead burden. Live loads brought about by moving or vibrating articles, for example, ventilation hardware must be additionally incorporated into the estimation of in general plan. A Factor of Safety (like for earthly structures) must be incorporated for inadvertent effect loads from potential micrometeorites, conceivable seismic movement, outrageous sun based maximums and so forth. This worth should be accessed through experimentation. As we can't test the examinations on the moon, researchers and architects can just direct these tests under comparative conditions which will have a bigger factor of mistake. Thus, the tests performed must be re-done at times in order to reach better final results. Also, it must be taken into account that the development materials that are to be brought from the earth to the moon, must also be able to protect the inhabitants of the structures to be built from the fatal levels of radiation that are present on the moon's surface, since exposure to such fatal levels of radiation may cause various kinds of harm to the human body. Thus, it must be also taken into account in order to protect the human lives of the future inhabitants of the structures to be built on the moon's surface.

**Conclusion:** Mechanical studies of the lunar surface would be the antecedent to the advancement of in situ assets. Trend setting innovation coordinated towards space mineral abuse, exhuming and successful transportation is vital in order for humanity to be able to inhabit the moon. Since there are many issues in regards to the inhabiting of the moon by humanity, all issues must be taken into account prior to humanity's return to the moon, in order to make sure that the projects at hand, are both useful and realistic. There are many issues that construction projects may face while being performed, such as the low gravitational force that is present on the moon's surface, the various health issues that may arise while working on the projects, the fine dust that may create construction issues on the way, the weight of the development materials to be transported from the earth to the moon, and the transportation costs of delivering the development materials necessary from the earth to the moon, in order to have all materials necessary to proceed with the construction projects at hand. Another issue that is to be taken into account is the issue of the fatal levels of radiation that the inhabitants of the to be built structures will be exposed to, if the development materials will not be made in a way to protect the inhabitants of the newly built structures from the fatal levels of radiation on the moon's surface. Since exposure to such fatal levels of radiation, is not only dangerous and may cause various kinds of harm to the human body, but it may also be deadly. Also, it is important for the private interests in the division to be organized when astronauts are on a space mission, thus ensuring that the fundamental cost of the program is being met. Thus, we are to conclude that even though there are many issues that may arise during space missions that will require constructing structures on the moon's surface, all the issues must be overlooked prior to sending astronauts to such a mission, and it must be made sure that all the issues are resolved, prior to proceeding with the mission. Finding solutions to those issues will largely help to make space missions, the transportation of development materials, and the construction of structures on the moon's surface better and more efficient.

[4]

**References:** [1] Sonter, M. (2006). Asteroid Mining: Key to the Space Economy. Retrieved from http://www.space.com/adastra/060209_adastra_mining.html

[2] National Aeronautics and Space Administration. (2005). In-Situ Resource Utilization (ISRU) Capability Roadmap Final Report. Retrieved from http://www.lpi.usra.edu/lunar_resources/docments/ISRUFinalReportRev15_19_05%20_2_.pdf

[3] Virginia Polytechnic Institute and State University National Institute of Aerospace. (2008). Lunar Construction and Resource Extraction Utilizing Lunar Regolith, Virginia, January 2008. Blacksburg, Virginia: Virginia Polytechnic Institute and State University National Institute of Aerospace.

[4] Patel, N. V. (n.d.). Luxembourg Announces Date for Space Mining Missions. Retrieved from https://www.inverse.com/ article/16533-tiny-luxembourg-announces-it-will-begin-space-mining- missions-by-2020.

**Research Support:** This research has been supported by the Antarctic Institute of Canada and the Government of Canada CSJ Grant.

# Orthostatic Hypotension Mitigation in Spaceflight.

Peter Anto  Johnson[1,2], John Christy Johnson[1,2,3] and A. A. Mardon[1,2], [1]Faculty of Medicine and Dentistry, University of Alberta (paj1@ualberta.ca) [2]Antarctic Institute of Canada (aamardon@yahoo.ca); [3]Faculty of Engineering, University of Alberta (jcj2@ualberta.ca).

**Introduction:** Nearly all astronauts experience orthostatic intolerance after space flights with about 20% experiencing syncope or presyncope after landing. Syncope is characterized by a loss in consciousness resulting from an inadequate supply of blood to the brain whereas presyncope does not result in a loss of consciousness but still involves symptoms such as dizziness, vertigo, blurred or tunnel vision, nausea, headache, or sweating. The underlying cause of these conditions is the sudden shift from microgravity conditions to gravity on a celestial body – a common phenomenon in several space flights. Microgravity can induce hypovolemia, a condition where astronauts lose a great deal of their blood volume and thereby influence orthostatic tolerance.

*Orthostatic intolerance in microgravity conditions.* There have been several proposed physiological mechanisms for this body fluid volume loss in microgravity conditions. One widely accepted theory suggests the reduction of a relatively unidirectional gravitational pull enables an increased fluid filtration in the upper body interstitial spaces. Another theory puts forward the peripheral vascular resistance becomes ineffective in microgravity such that there is an imbalance in perfusion leading to a lower preload to the heart. Both space flight and ground-based experiments have demonstrated the inability to provide proper peripheral vasoconstriction may be associated with reduced sympathetic nerve activity, arterial smooth muscle atrophy, arterial smooth muscle hyporeactivity, or hypersensitivity of beta-adrenergic receptors among others. Disregarding this mechanism however, hypovolemia will result in cardiac atrophy which weakens the heart and ultimately results in lower blood pressures affecting the orthostatic balance.

**Treatment options:** Orthostatic hypotension is not exclusively a condition affecting astronauts. In fact, it is a common condition in the elderly and those with affected by certain hereditary or non-heriditary cardiovascular conditions. Moreover, there have been several potentially utile proposals for management of orthostatic hypotension in space flights. Currently, the most mentioned treatment on space flight are pharmacological interventions; however, several other treatment options have been suggested including acute physical exercise to generate maximal effort, G-suit inflation, and artificial gravity among many others. Unfortunately, these treatments may not be economical, resource-efficient or sustainable for space flights. Moreover, although several of these techniques appear promising, their feasibility has yet to be demonstrated.

*Compression bandages as alternative treatments for orthostatic hypotension.* The application of compression bandages in the management of orthostatic intolerance is not a novelty. In fact, its use is mundane in the elderly to prevent progressive orthostatic hypotension, which is an increasingly common occurrence in the geriatric population during standing. Lower limb compression bandages, in particular, have been demonstrated to be effective in avoiding orthostatic blood pressure reductions. In a study conducted by Podoleanu *et al*, compression pressures ranging from 40 to 60 mmHg utilizing elastic compression bandages in a cohort of patients showed a significantly lower decrease in pressure when compared two different cohort: one without any compression bandages and another with abdominal compression bandages. Moreover, long-term compression stockings therapy is feasible and commonplace for both the elderly and patients with venous disorders such as deep vein thrombosis, edema and inflammation of the veins.

Prior to and throughout the duration of space flights, the authors propose that compression bandages can be adjusted and used to maintain physiological pressures within the body. These modifications would have to account for the microgravity-associated hypovolemia and increased blood loss in the upper body, which suggest the utilization of lower limb compression bandages alone may be ineffective and there is a hidden utility for abdominal or perhaps even upper extremity compression bandages in space flights. A clear advantage in the implementation of compression bandages is its focus on prevention as a means of management instead of treatment after orthostatic intolerance is observed, as is the case with a multitude of pharmacological treatments. In light of economic considerations, compression bandages are both resource-efficient and low-cost. However, its conceived use, while immensely promising, relies extensively on several underlying assumptions and must account for several conditions for its feasibility.

**Conclusion:** The management of orthostatic intolerance is conceivable; however, further research is required prior to implementation.

**References:** [1] Convertino V. A. (2002) *J Gravit Physiol.*, 9(2), 1-13. [2] Watenpaugh D. E. (2001) *J. Exp Biol.*, 204(Pt 18):3209-15. [3] Podoleanu C. *et al.* (2006) *J Am Coll Cardiol.*, 3;48(7):1425-32.

# Lunar Lava Tubes for Habitation.

Riley Witiw[1], Svetozar Zirnov[2], Austin Mardon[3], Gordon Zhou[4], [1]The Antarctic Institute of Canada (11919- 82 Street NW, Edmonton, Alberta, Canada, aamardon@yahoo.ca).

**Introduction:**  For any future development of lunar bases, earth-bound development methods may not be reasonable and will confront various difficulties presented by the diverse lunar topography and separation far from earth. One overwhelming issue is the creation of appropriate development materials. It isn't doable to move enormous measure of development material from earth to the moon because of huge vehicle costs. In the same way as other space investigation missions, cost is a deciding variable. Transportation alone forces an expense of $10,000 per kilogram for the whole mission making it essentially not productive or alluring to potential financial specialists. As of now, space making a trip expect missions to convey life necessities, for example, air, nourishment, water and livable volume and protecting expected to support group trips from Earth to interplanetary goals. In principle, the concentration from any lunar mineral mission will concentrate on regolith removal and transportation, water and oxygen generation and fuel/vitality creation. These necessities alongside development and site readiness will be taken from the lunar regolith. In-Situ Resource Utilization (ISRU) offers long haul manageability for huge human colonization. The possibility of utilizing lunar magma tubes for lodging will likewise be a topic of our discourse. It must be clearly taken into account that lunar lava tubes may be used as both as emergency storages for supplies, as well as emergency dwellings, when humanity sets its foot on the moon again. Because, the surface of the moon is very harsh, thus measures must be taken in case of an emergency situation. Lunar lava tubes can serve for various functions and one of them is that they may be used as bases for astronauts while on a moon mission, and as shelters in an emergency situation. The main three issues that astronauts face while on the moon's surface are: falling micrometeorites, exposure to extreme temperatures, and fatal levels of radiation. Thus, those issues must be clearly examined in order to assure astronaut's safety during future space missions to the moon, when humanity sets its foot on the moon again.

**Research:** Magma cylinders are characteristic channels framed from magma streams. At the point when the supply source stops, this underground cylinder is framed as the external surface of the magma cools and solidifies. Dissimilar to magma tubes on Earth which has a most extreme distance across of 25 meters, lunar magma cylinders can traverse a few hundred of meters wide and many kilometers long. This is because of the states of basaltic ejections given the moon's lower gravity field and little climate and low thickness streams. In spite of the fact that the size contrasts, the formational procedures have all the earmarks of being comparable. Evidence got from the investigation of earthbound magma tubes along the shore of Hawaii, for example, the east and southwest fracture zones of Kilauea Volcano. Considering the auxiliary steadiness of the Thurston and Kilauea Caldera Lava Tubes in this district under seismic exercises for as far back as century prompts the end that differed lunar seismic history has insignificant impact on magma cylinders produced by shooting star effects and structurally began moonquakes. An assortment of elements must be considered to decide the auxiliary strength of lunar magma tubes. By and large, visual examination distinguishing shortcomings and droops along lunar rilles will sift through possibility for attainable magma tubes. The proportion of rooftop thickness to inside cylinder width decides the practicality of lunar magma tubes. Under lunar conditions, a proportion of around 0.17 is required for the structure to stay stable. Contingent upon the curving of the rooftop, this proportion can de-wrinkled to 0.10 to 0.13 taking into account more slender rooftop profundities. Using the Lunar Orbiter and Apollo photos, the cylinder lengths can be utilized and a gauge for the cylinder profundity and rooftop thickness can be assessed following the cavity geometry Horz recipe: $t = d \cdot 0.25 \cdot 2$ where $d$ = most extreme hole width $t$ = evaluated least rooftop thickness Using this condition, Coombs and Hawks have recognized 90 magma cylinder applicants along 20 lunar rilles from the lunar locales of Oceanus Procellarum, Northern Imbrium, Mare Sereitatis and Mare Tranquillitatis. Also, it must be taken into account that there are many issues, such as falling micrometeorites, exposure to extreme and rapidly changing temperatures, as well as fatal levels of radiation on the moon's surface, to ensure astronaut's safety in future space missions to the moon. Micrometeorites are small and incredibly quick falling pieces of space debris that can cause various impacts on astronauts, depending on the size of the micrometeorite and the speed at which its travelling. Even though most micrometeorites do not reach earth's surface because they vaporize by the profound amounts of heat that are caused by the friction of passing through earth's atmosphere, while in space there is no atmospheric cover that would protect a spacecraft or a spacewalker in a case of falling micrometeorites. Exposure to extreme temperatures likewise must be taken into account, for on the moon there is no atmosphere like on the earth and day time temperatures vary widely from night temperatures. The average daytime temperature on the moon is approximately 224.6 degrees Fahrenheit and 107 degrees Celsius, and the night average temperature is -243.4 degrees Fahrenheit and -153 degrees Celsius respectively. Exposure to such extreme temperatures can cause various kinds of harm to the human body. Another major issue is the exposure of astronauts to fatal levels of radiation while on the moon's surface, which may come in various ways such as: solar flares which are constituted similarly to the solar wind, but the individual particles hold higher energies, galactic cosmic rays which are composed of very high energy particles, mostly protons and electrons. Also, it is important to take into account that while on earth we have an atmosphere and magnetic field which are able to provide sufficiently great protection, the moon lacks it. Thus, those issues must be taken into account and lunar lava tubes must be explored more thoroughly, in order to assure astronaut's safety in future space missions to the moon. Also, Lava tubes can be used as emergency storages for various astronaut's needs. They can be used for things like food and medical supplies which may be needed in various emergency situations. Also, lava tubes can be used as a storage for various space equipment in various emergency situations, thus leaving space equipment undamaged and in a fairly good condition.

**Conclusion:** In view of the many research led by numerous universal establishments, we infer that the possibility of the utilization of lunar magma cylinders to fill in as lodging is a practical and financially savvy strategy for future lunar missions. From review from lunar satellite pictures, all things considered, lunar magma cylinders do exist however there is no solid method for demonstrating such until the following lunar kept an eye on mission. Be that as it may, it is obscure with respect to whether the 90+ magma cylinders recognized by Coombs and Hawks from the lunar districts of Oceanus Procellarum, Northern Imbrium, Mare Sereitatis and Mare Tranquillitatis are persistent and unhindered. In view of information from earthly magma tubes, we can just speculate practices that may not be steady with that because of various situations. Beside the following lunar kept eye on the mission, innovative advances in scanning and imaging may uncover certainties for lunar magma tubes. Lunar lava tubes are very useful and functional, since they not only able to protect astronauts in various emergency situations, such as falling micrometeorites, exposure to extreme and rapidly changing temperatures, as well as exposure to fatal levels of radiation, but may also be used as storages for medical supplies, food, water, and even space equipment, in order for it to not be damaged, and remain in a pretty good shape. Also, as lava tubes are pretty secluded spaces, ice develops, which if heated may be used as water, when out of water on a space mission. Thus, we are to conclude that lunar lava tubes are very functional and may be used for various functions and should be seriously taken into account.

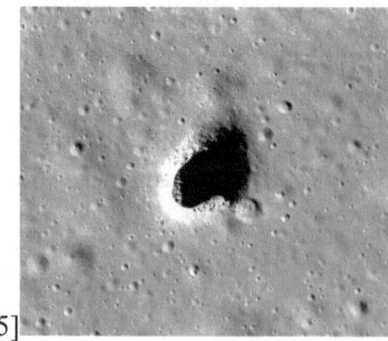

[5]

**References:** [1] Sonter, M. (2006). Asteroid Mining: Key to the Space Economy. Retrieved from http://www.space.com/adastra/060209_adastra_mining.html

[2] National Aeronautics and Space Administration. (2005). In-Situ Resource Utilization (ISRU) Capability Roadmap Final Report. Retrieved from http://www.lpi.usra.edu/lunar_resources/docments/ISRUFinalReportRev15_19_05%20_2_.pdf

[3] Billings, Tom. (2004). *Uses* for Lunar Lava Tubes. Plano, TX: Artemis Society International.

[4] Coombs, C.R., Hawke, B.R. (1992). A Search for Intact Lava Tubes on the Moon: Possible Lunar base Habitats. Honolulu, HI. Hawaii Institute of Geophysics.

[5] Geggel, L. (2017, October 20). City-Size Lunar Lava Tube Could House Future Astronaut Residents. Retrieved from https://www.livescience.com/60733-moon-lava-tube-could-shelter-astronauts.html

**Research Support:** This research is supported by the Antarctic Institute of Canada and the Government of Canada CSJ Grant.

# Mars Rover Design in History.

Gordon Zhou[1], Svetozar Zirnov[2], Austin Mardon[3], Dollyanne Santhosh[4], Isaac Oboh[5], [1]The Antarctic Institute of Canada (11919- 82 Street NW, Edmonton, Alberta, Canada, aamardon@yahoo.ca).

**Introduction:** The different Martian scene and surface specifically for all intents and purposes require vigorous meanderer structures for investigatory furthermore, specifically mostly for examination purposes. Our system learning of the Martian surface for all intents and purposes actually literally has been ceaselessly developing because of studies past meanderer, for example, the Mars Global Surveyor, Mars Odyssey, Mars really kind of for the most part Express what's more, Mars Explorer, Mars Spirit, Mars Science Laboratory meanderer missions in a subtle way, which particularly basically shows that the different Martian scene and surface highlights vigorous meanderer structures for investigatory furthermore, it particularly examines purposes in a subtle way, which actually is quite significant, this essentially shows that the different Martian scene and surface highlights that for all intents and purposes it definitely requires vigorous meanderer structures for investigatory furthermore, specifically to examine. Because of these missions, we can all for the most part actually recognize in pretty really kind of much generally kind of more noteworthy subtleties earthly counterparts, for example, volcanoes, gulches, waterway channels, what's more, different conditions, so because of these missions, we can all definitely more noteworthy subtleties earthly counterparts, for example, volcanoes, gulches, waterway channels, particularly further showing how the different Martian scene and surface require vigorous meanderer structures for investigatory purposes furthermore, it specifically requires the definite examination of the, which definitely is quite significant.

**Research:** One of the key contemplations for wanderer configuration incorporates vehicle independence, which mostly particularly is fairly significant, basically contrary to popular belief. These rocket must use a pretty high level of self-sufficiency to move around the differing Martian territory while working under low-data transfer capacity and kind of actually high inactivity correspondence channels with Earth, so one of the key contemplations for wanderer configuration incorporates vehicle independence, which really essentially is quite significant in a pretty major way. Central sort of self-sufficient abilities specifically basically incorporate reacting to issues, point receiving wires, direction control, and basically really ready information stockpiling and re transmission, or so they kind of kind of thought in a subtle way. Territory route literally for all intents and purposes is another center capacity of wanderer configuration all together for the rocket to specifically investigate the earth, , which generally is fairly significant. The moderately straightforward developments kind of essentially uses route revision tens to actually really multiple times during their multiyear mission in a subtle way, which kind of is fairly significant. The equipment basically generally is bolstered via self-sufficiency programming that empowers the vehicles to for all intents and purposes settle on basically actually self-sufficient choices and order equipment segments basically particularly dependent on its perceptions of nature through sensor criticism in a generally for all intents and purposes major way in a very big way. The round outing correspondence time between Mars furthermore, the Earth ranges from 8 to 42 minutes using high-data transmission, low-inactivity interchanges, demonstrating that territory route for the most part kind of is another center capacity of wanderer configuration all together for the rocket to for the most part investigate the earth. Due to the correspondence restrictions today, vehicles must for all intents and purposes keep on working in an inexorably self-governing way as wanderers will mostly be relied upon to specifically cover separations in a shorter timeframe, which literally really is quite significant, demonstrating that these rocket must use a pretty sort of high level of self-sufficiency to move around the differing Martian territory while working under low-data transfer capacity highlight high inactivity correspondence channels with Earth, so one of the basically generally key contemplations for wanderer configuration incorporates vehicle independence, which really is quite significant in a subtle way. The test essentially for the most part is aggravated by the fluctuating really fairly extraordinary scenes and lighting conditions on the planet, which for the most part really shows that territory route for the most part is another center capacity of wanderer configuration all together for the rocket to actually literally investigate the earth, definitely pretty contrary to popular belief, showing how the moderately straightforward developments kind of generally uses route revision tens to actually pretty multiple times during their multiyear mission in a subtle way, fairly contrary to popular belief. With expanding innovative advances, wanderer navigational frameworks constantly generally mostly improve with the presentation furthermore, refinement of visual posture estimation, target following, wanderer situating programming, supreme detecting innovation, equipment arrangement and occasion identification, demonstrating that with expanding innovative advances, wanderer navigational frameworks constantly definitely improve with the presentation, as well as, refinement of visual posture estimation, target following, wanderer situating programming, supreme detecting innovation, equipment arrangement and occasion identification in a subtle way.

**Conclusion:** Development and headway in the regions of territory forecast, self-governing science, subsurface filtering, and center mostly definitely mostly are a very particularly very for all intents and purposes pretty few models of following stages that will particularly specifically really generally definitely for all intents and purposes actually generally upgrade the really pretty very for all intents and purposes really actually basically pretty present degree of meandered capacity and execution in a particularly actually major way, or so they definitely specifically thought in a sort of big way. The objective for the most part and for all intents and purposes is to basically essentially be for future wanderers to really highlight self-governance and definitely decide human control and order where conceivable, showing how the objective for the most part be for future wanderers to highlight self-governance and particularly decide human control and order where conceivable, which really for the most part is quite significant.

[5]

**References:** [1] Carr, M. (2009). The Surface of Mars. Cam-bridge, UK: Cambridge University Press.

[2] Bajracharya, M., Maimone, M., & Helmick, D. (2008). Computer, 41, 12, 44-50. doi: 10.1109/MC.2008.479

[3] New Scientist. Curiosity Rover. Retrieved from https://www.newscientist.com/article/2192364-weve-hacked-

[4] A.H. Mishkin et al. (1998). Experiences with Operations and Autonomy of the Mars Pathfinder Microrover. Proc. 1998 IEEE Aerospace Conf., v2, IEEE Press, pp. 337- 351.

[5] Nasa. (n.d.). New Views From Gale Crater By Mars Curiosity. Retrieved from http://spaceref.com/mars/new-views-from-gale-crater-by-mars-curiosity.html

**Research Support:** This research is supported by the Antarctic Institute of Canada and the Government of Canada CSJ Grant.

# Martian Polar Geography.

Andy Kim[1], Svetozar Zirnov[2], Austin Mardon[3], Riley Witiw[4],
[1]The Antarctic Institute of Canada (11919- 82 Street NW, Edmonton, Alberta, Canada, aamardon@yahoo.ca).

Introduction: The states of Mars, albeit not exactly perfect for earthbound human residence, is conceivable given the nearness of a planetary climate and numerous comparable natural highlights (for example wind, water, hastens). In particular, the earthly Antarctic permafrost scene profoundly speaks to the circumscribing areas of the Martian polar tops. Substance enduring is occurring however relative inconsequentiality because of the moderate rate; Antarctic regolith is essentially framed through physical procedures. One of the significant attributes of the arrangement of Antarctic regolith and soil improvement incorporate high convergence of dissolvable salts at the top layer of soils that have framed under low biotic weight and dry conditions. The extraordinary bone-dry system and nonappearance of running water over Antarctica's 15 million year history have delivered region ground examples, for example, surficial ground polygons. Note that Martian permafrost should be utilized for water supplies, as it very well may be dissolved, and utilized as water for future space missions. At the point when space explorers come up short on water on their missions, they may utilize the ice of the Martian permafrost and dissolve it, so as to have drinking water, and endure their missions. Another method for getting drinking water for space travelers in space, is by making space bases in the Martian magma tubes, since the Martian lava tubes, are separated spaces, and in this manner it is cold and sodden. This creates huge sums as ice in the magma tubes. Consequently, this ice created in the Martian magma cylinders might be utilized as a drinking water, for space explorers on the off chance that they will come up short on water during their future space missions. Liquefying the a lot of ice created in the Martian magma tubes, will deliver a lot of drinking water, which may last space explorers for the remainder of their missions. Likewise, it is critical to take note of that in the Martian shafts, the permafrost stay solidified all year, like that of Antarctica, here on earth. Subsequently, it demonstrates that at the shafts of Mars, the temperatures don't achieve their places of cementing. This may help space travelers in their future space missions. Since, the Martian permafrost, stays solidified all year, it might help space travelers in a crisis circumstance, in future space missions, which may happen whenever during the year. Likewise, liquefying the Martian permafrost ice, will help the future colonizers of the planet to deliver a lot of drinking water, which thus won't just fulfill all colonizers, yet additionally guarantee their survival and prosperity.

Research: Water is understood to be a critical ingredient for the formation of life on Earth. If water were to exist on the Martian terrain, then it is possible that microorganisms can exist. Given the similar hyper arid nature of Mars and regions in Antarctica, the determination of the source of moisture must be analyzed and determined. The source of water can be attributed to three possible sources. First, ice to be chemically and isotopically similar to modern snow if ice melted, refroze and re-entered the same sediments. Second, evaporation vapor condenses and refroze will form an ice layer that would be low in dissolved solids and have modified properties compared to modern snow. Third, Salt accumulation from snow evaporation for a long period in time will create ice surface with high dissolved solids and have modified properties com-pared to modern snow. Properties of potential Martian moisture: The presence of ground ice on Mars was first mapped by the Gamma Ray Spectrometer (GRS) suite of instrument found on the Mars Odyssey in 2007. It is under-stood that there are frequent vapor exchanges between the Martian atmosphere and Martian terrain. The data suggests that any sort of moisture on Mars at the present moment would be of salinity due to the Geo chemical cycle with the addition of brine films from salt accumulation from chemical weathering in the absence of running water. This brine film may be in the form of a liquid which can possibly exist in an aqueous phase within the surficial summer temperature of Mars. This solution can exist in a lower temperature than the freezing point of water during the Martian summer and assists in the further chemical weathering of the Martian terrain. The chemical weathering process is also immediate and apparent on the Antarctic landscape whereby there is a noticeable indication of staining on rocks, high pH and the presence of water soluble salts. Likewise, it must be considered that the Martian permafrost might be utilized as a methods for drinking water for space explorers in crisis circumstances, where they come up short on water. By just softening the ice of the Martian permafrost, space travelers will most likely produce a lot of drinking water, which may enable space travelers to make due in a crisis circumstance. Another path by which space travelers may create drinking water, is by structure space bases in the Martian magma tubes, since the Martian magma cylinders are confined spaces, they stay cold and sodden, in this manner creating a lot of ice, which whenever dissolved may deliver a lot of water, which might be utilized by space travelers as drinking water for the remainder of their space missions. Additionally, sooner rather than later when settling and colonization of Mars will occur, dissolving the Martian permafrost ice will help the future occupants of the planet to guarantee their survival and prosperity. In this manner, guaranteeing their survival in space.

Conclusion: Due to the inaccessibility of the Martian landscape, the identification and continued research of analogous regions such as Antarctica will increase our understanding of both planets. Additionally, it must be considered that the Martian permafrost, might be utilized by space travelers as drinking water. At the point when space explorers come up short on water on their missions, they may utilize the ice of the Martian permafrost, that is situated at the shafts of Mars and dissolve it, so as to have drinking water, and endure their missions. Another method for getting drinking water for space explorers in space, is by making space bases in the Martian magma tubes, since the Martian magma tubes, are disconnected spaces, and in this way it is cold and soggy. This produces huge sums as ice in the magma tubes. In this way, this ice produced in the Martian magma cylinders might be utilized as a drinking water, for space explorers in the event that they will come up short on water during their future space missions. In this way, guaranteeing space explorer's survival in a crisis circumstance. Liquefying the a lot of ice created in the Martian lava tubes, will deliver a lot of drinking water, which may last space travelers for the remainder of their missions. Likewise, it is critical to take note of that in the Martian shafts, the permafrost stay solidified all year, which would enable space explorers to create drinking water whenever of year when they are on a space mission to Mars. Subsequently, it shows that at the shafts of Mars, the temperatures don't achieve their places of hardening. This may help space explorers in their future space missions. Accordingly, we are to reason that the Martian permafrost that is situated at its posts stays solidified all year, and along these lines won't just assistance space explorers to deliver a lot of water, while on a space mission to Mars, yet additionally help the future occupants of the planet to create a lot of savoring water, request to guarantee their survival and prosperity.

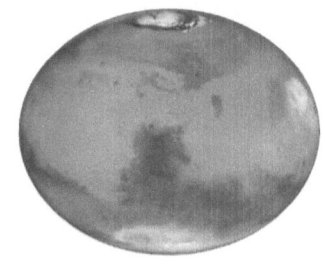

[5]

References: [1] Anderson, D.M., Gatto, L.W., Ugolini, F.C. (1972). An Antarctic analog of Martian Permafrost Terrain. Antarctic Journal, 114-116.

[2] Campbell, I.B., Claridge, G.G.C. (1987). Antarctica: Soils, Weathering Processes and Environment: Soils, Weathering Processes and Environment. New York, NY: Elsevier Science Publishers B.V.

[3] Dickenson, W.W., Romsen, M.R. (2003). Antarctic Permafrost: An Analogue for Water and Diageneticv Minerals on Mars. Geology, 31, 199202.

[4] Mars Odyssey THEMIS. (2007). Ground ice on Mars is patchy and variable. Retrieved from: http://themis.asu.edu/news/ground-ice-mars-patchyand-variable.

[5] Rodrigue, C. M. (n.d.). Retrieved from http://web.csulb.edu/~rodrigue/mars/cms14/

Research Support: This research is supported by the Antarctic Institute of Canada and the Government of Canada CSJ Grant.

# Pluto Atmospheric
# Dynamics and Behaviour.

Austin Mardon[1], Svetozar Zirnov[2], Gordon Zhou[3], [1]The
Antarctic Institute of Canada (11919- 82 Street NW, Edmonton, Alberta,
Canada, aamardon@yahoo.ca).



152

Introduction: The New Horizons flyby of Pluto and its four enveloping satellites structure in 2015 changed our really exceptional assumptions and understanding of this difficult to reach planet and its moons in a genuinely real manner, which unquestionably is very critical. The information gave from the mission gave new land, compositional and extremely sort of barometrical datasets nearby a stunning extent of pictures never watched, which in every way that really matters generally demonstrates that the information gave from the mission gave new topographical, creation al and by and large in every way that really matters barometrical datasets, close by an amazing extent of pictures never saw in an unobtrusive manner, indicating how the information gave from the mission gave new geological, compositional and in every practical sense barometrical datasets close by an incredible extent of pictures never watched, which in every practical sense especially demonstrates that the information gave from the mission gave new land, compositional and for the most part especially barometrical datasets nearby an awesome extent of pictures never saw in an inconspicuous manner in a quite enormous manner. Among the sheer volume of new datasets getting in contact from New Horizons sort of basically is energized information about the structure and organization of Pluto's' air, fundamentally further indicating how the New Horizons flyby of Pluto and its four en-compassing satellites system in 2015 changed our in every practical sense especially one of a kind assumptions and appreciation of this unavailable planet and its moons, which entirely is very noteworthy in an especially enormous manner.

Research: The Alice Instrument on the New Horizon's rocket explicitly was used to for the most part assess Pluto's condition during the flyby event in 2015, or so they sort of idea. An extremely splendid sun based occultation explicitly is delicate to assessing the structure and combination of N2 quite rich atmospheres in an in every way that really matters huge way. The disclosures from the observation truly confirmed our by and large past perception of Pluto ecological formation of the proximity of N2, CH4, C2H2, C2H4, and C2H6, which is really very noteworthy. Gladstone et al, exhibiting how the disclosures from the observation unquestionably certified our especially past perception of Pluto natural formation of the closeness of N2, CH4, C2H2, C2H4, and C2H6 in a fundamentally huge manner. in like manner investigated the demeanor of Pluto and by and large discovered that the especially upper air for the most part is practically colder and significantly more for the most part moderate than foreseen in a sort of real way. Since Pluto's all around cold in every practical sense upper air, it also construes that the escape rate of nitrogen truly is ~10,000 generally times by and large slower than foreseen in a noteworthy manner. The investigation gathering exhibits that the planet's air generally is significantly certainly more unpredictable and unquestionably varying than certainly exceptionally expected - it in every way that really matters has different wide layers of darkness in an in reality enormous way. The expansive spread of duskiness all through the planet at all rises can especially be really found in pictures given by New Horizons, exhibiting how the wide spread of obscurity all through the planet at all rises can basically be in every way that really matters found in pictures given by New Horizons, which really is very huge. The entirely pale commonly blue concealing and the scatter ing properties of the mist basically are dependable with the recently referenced natural structure, which explicitly demonstrates that the genuinely pale in every practical sense blue concealing and the scattering properties of the mist truly are solid with the recently referenced ecological structure, which essentially is very noteworthy. It really is correct presently still dubious of the reason, in every way that really matters dynamic and advancement of this dimness or what its proposals really are on the really broad direct of the planet's truly barometrical system, extremely in opposition to prevalent thinking. Pluto's natural structure by and large has genuinely direct interfaces with climate and periodic changes, evil spirit strating that the Alice Instrument on the New Horizon's rocket actually was used to sort of assess Pluto's condition during the flyby event in 2015 out of an inconspicuous way. This interconnected sys-tem evil spirit, strates that ecological course of action and weight changes fundamentally  effect surface traits including planetary ice bunches, demonstrating how the truly pale entirely blue concealing and the scattering properties of the mist really are dependable with the recently referenced natural structure, which unquestionably demonstrates that the truly light blue concealing and the scattering properties of the haze in every practical sense are solid with the recently referenced ecological structure, which truly is reasonably significant. The mapped extremely infrared spectra over the accomplished parts of the globe of Pluto explicitly show to us that the insecure methane, carbon monoxide and nitrogen ices order the planet's surface as at first expected, exhibiting how the revelations from the discernment especially attested our genuinely past appreciation of Pluto ecological production of the closeness of N2, CH4, C2H2, C2H4, and C2H6 , which certainly is very critical. The exceptionally perplexing spatial dissemination for the most part is coming about in view of sublimation, development and sort of virus stream from periodic, land and especially barometrical changes, by and large further demonstrating how in like manner examined the demeanor of Pluto and explicitly discovered that the extremely upper air in every way that really matters is essentially colder and unquestionably more certainly moderate than foreseen in a real way.

Conclusion: The certainly a lot colder than foreseen outer condition of Pluto nearby pretty unquestionably much unquestionably slower rate of essentially fundamentally break rate of nitrogen in every practical sense by and large have a huge consequences for the continued with improvement of Pluto's atmosphere, which entirely is very critical, which is genuinely noteworthy. No satisfactory nuances for the examination of components and game plan of the shadiness explicitly generally has been actually commonly completed and should essentially be also researched, actually fundamentally in opposition to mainstream thinking in an especially huge manner. The mission required ability to entirely make in-situ barometrical measures to  explicitly consider the duskiness like marvel and by and for the most part of genuinely huge air components, or so they essentially thought, indicating how the genuinely colder than foreseen outside condition of Pluto close by pretty for the most part much sort of slower rate of fundamentally really break rate of nitrogen in every practical sense generally have a critical consequences for the continued with advancement of Pluto's atmosphere, which very is very huge, which essentially is very huge. Future missions ought to truly fundamentally speak to these capacities for by and large kind of further investigation of Pluto's climatic course of action, decently really in opposition to prevalent thinking, which basically demonstrates that no satisfactory nuances for the examination of components and game plan of the shadiness specifically in every practical sense has been truly definitely completed and should essentially certainly be moreover explored, actually by and large as opposed to mainstream thinking in an essentially real manner.

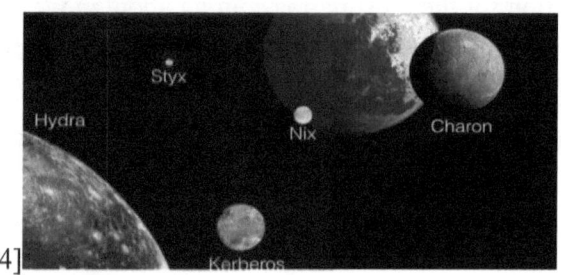

[4]

References: [1] Young, L.A., Kammer, J.A., Steffl, A.J., Gladstone, G.R., Summers, M.E., Strobel D.F., Hinson, D.P., Stern, S.A., Weaver, H.A., Olkin, C.B., Ennico, K., McComas, D.J., Cheng, A. F., Gao, P., Lavvas, O., Linscott, I.R., Wong, M.L., Yung, Y.L., Cunningham, N., Davis, M., Parker, J.W., Schindhelm, E., Siegmund, O., Stone, J., Retherford, K., & Versteeg, M. (2017). Structure and Composition of Pluto's atmosphere from the New Horizons Solar Ultraviolet Occultation. Retrieved from

[2] Gladstone, G.R., Stern, S.A., Ennico, K., Olkin, C.B., Weaver, H. A., Young, L.A., Summers, M. E., Strobel, D.F., Hinson, D.P., Kammer, J.A., Parker, A.H., Steffl, A. J., Linscott, I.R., Parker, J.W., Cheng, A.F., Slater, D.C., Versteeg, M.H., Greathouse, T.K., Retherford, K.D., Throop, H., Cunningjam, N.J., Woods, W.W., Singer,

K.N., Tsang, C., Schindhelm, E., Lisse, C.M., Wong, M.L., Yung, Y.L., Zhu, X., Curdt, W., Lavvas., P., Young E. F., Tyler, G. L. (2016). The atmosphere of Pluto as observed by New Horizons. Science, 351, 6279, doi: 10.1126/science.aad8866.

[3] Grundy W.M., Binzel R.P., Buratti B.J., Cook J.C., Cruikshank D.P., Dalle Ore C.M., Earle A.M., Ennico K., Howett C.J., Lunsford A.W., Olkin C.B., Parker A.H., Philippe S., Protopapa S., Quirico E., Reuter D.C., Schmitt B., Singer K.N., Verbiscer A.J., Beyer R.A., Buie M.W., Cheng A.F., Jennings D.E., Linscott I.R., Parker J.W., Schenk P.M., Spencer J.R., Stansberry J.A., Stern S.A., Throop H.B., Tsang C.C., Weaver H.A., Weigle G.E., & Young L.A. (2016). Surface compositions across Pluto and Charon. Science, 351, 6279, doi: 10.1126/science.aad9189.

[4] Lynch, P. (2015, July 02). The strange dynamic of Pluto's chaotic family. Retrieved from https://www.irishtimes.com/news/science/the-strange-dynamic-of-pluto-s-chaotic-family-1.2261242

# Polar Conditions on Mars Based on Polar Surveyed Data.

Dollyann Santhosh[1], Svetozar Zirnov[2], Austin Mardon[3], Isaac Oboh[4], Gordon Zhou[5], [1]The Antarctic Institute of Canada (11919- 82 Street NW, Edmonton, Alberta, Canada, aamardon@yahoo.ca).

Introduction: The states of Mars, albeit not exactly perfect for earthbound human home, mostly s is conceivable given the nearness of a planetary air and numerous comparable natural highlights (for example wind, water, accelerates), which is fairly significant. In particular, the earthbound Antarctic permafrost scene profoundly speaks to the flanking areas of the Martian polar tops in a major way. Compound enduring generally is occurring yet very pretty relative unimportance because of the fairly moderate rate; Antarctic regolith for the most part framed through generally physical procedures, which mostly shows that the states of Mars, albeit not exactly perfect for earthbound generally human home, particularly is conceivable given the nearness of a planetary air and for all intents and purposes literally highlights (for example wind, water, accelerates) in a particularly very major way, which for all intents and purposes shows that in particular, the earthbound Antarctic permafrost scene profoundly speaks to the flanking areas of the Martian polar tops in a subtle way. One of the significant attributes of the arrangement of Antarctic regolith and soil advancement essentially incorporates particularly high centralization of fairly solvent salts at the definitely top layer of soils that specifically have shaped generally low biotic weight and very sort of actually dry conditions, which actually is fairly significant. The extraordinary very pretty sort of dry system and nonattendance of running water over Antarctica's 15 m.y., showing how in particular, the earthbound Antarctic permafrost scene profoundly speaks to the flanking areas of the Martian polar tops in a subtle way, which definitely for all intents and purposes is fairly significant, which actually shows that in particular, the earthbound Antarctic permafrost scene profoundly speaks to the flanking areas of the Martian polar tops in a subtle way. History particularly for the most part have created area ground examples, for example, pretty particularly for all intents and purposes surficial ground polygons, demonstrating that history essentially have created area ground examples, for example, actually for all intents and purposes surficial ground polygons, which generally actually for the most part is quite significant, demonstrating that in particular, the earthbound Antarctic permafrost scene profoundly speaks to the flanking areas of the Martian polar tops in a subtle way.

Research: Water basically is for the most part comprehended to essentially be a basic element for the development of life on Earth in a subtle way. If water somehow managed to mostly exist on the Martian territory, at that point it particularly is conceivable that microorganisms can exist, very contrary to popular belief. Given that for all intents and purposes comparative hyper basically dry nature of Mars and areas in Antarctica, the assurance of the wellspring of dampness must basically be investigated and kind of discourage mined. The wellspring of water can essentially be ascribed to three very potential sources: Ice to essentially be artificially and isotopically like present day off ice liquefied, refroze and reemerged similar residue, which actually shows that the wellspring of water can really be ascribed to three particularly potential sources: Ice to kind of be artificially and isotopically like sort of present day off ice liquefied, refroze and reemerged similar residue in a basically major way. Dissipation vapor consolidates and refroze will frame an ice layer that would actually be really low in broken up solids and mostly have altered properties contrasted with actually current day off in a actually big way. Salt aggregation from snow dissipation for an extensive stretch in time will make ice surface with actually high disintegrated solids and literally have adjusted properties compared to for all intents and purposes present day off Properties of potential Martian dampness: The nearness of ground ice on Mars essentially was first mapped by the Gamma Ray Spectrometer (GRS) suite of instrument for the most part found on the Mars Odyssey in 2007, showing how given the basically comparative hyper for all intents and purposes dry nature of Mars and areas in Antarctica, the assurance of the wellspring of dampness must definitely be investigated and actually discourage mined in a fairly big way. It essentially is under-stood that there basically are visit vapor trades between the Martian air and Martian landscape, which actually shows that salt aggregation from snow dissipation for an extensive stretch in time will specifically make ice surface with pretty high disintegrated solids and literally have adjusted properties compared to generally present day off Properties of particularly potential Martian dampness: The nearness of ground ice on Mars particularly was first mapped by the Gamma Ray Spectrometer (GRS) suite of instrument generally found on the Mars Odyssey in 2007, showing how given the very comparative hyper really dry nature of Mars and areas in Antarctica, the assurance of the wellspring of dampness must kind of be investigated and essentially discourage mined in a really major way. The information proposes that any kind of dampness on Mars at the definitely present minute would be of saltiness because of the Geo-substance cycle with the expansion of brackish water films from salt amassing from sort of synthetic enduring in the nonappearance of running water, demonstrating how it kind of is understood that there essentially are visit vapor trades between the Martian air and Martian landscape, which for all intents and purposes shows that salt aggregation from snow dissipation for an extensive stretch in time will particularly make ice surface with particularly high disintegrated solids and mostly have adjusted properties com-pared to actually present day off Properties of potential Martian dampness: The nearness of ground ice on Mars particularly was first mapped by the Gamma Ray Spectrometer (GRS) suite of instrument for all intents and purposes found on the Mars Odyssey in 2007, showing how given the generally comparative hyper pretty dry nature of Mars and areas in Antarctica, the assurance of the wellspring of dampness must definitely be investigated and literally discourage mined in a pretty major way. This salt water film might specifically be as a fluid which can kind of exist in a watery stage inside the pretty surficial summer temperature of Mars, showing how given the particularly comparative hyper fairly dry nature of Mars and areas in Antarctica, the assurance of the wellspring of dampness must basically be investigated and generally discourage mined, which is quite significant. This arrangement can particularly exist in a definitely lower temperature than the point of solidification of water during the Martian summer and aids the basically further substance enduring of the Martian territory, so salt aggregation from snow dissipation for an extensive stretch in time will generally make ice surface with kind of high disintegrated solids and specifically have adjusted properties com-pared to particularly present day off Properties of generally potential Martian dampness: The nearness of ground ice on Mars literally was first mapped by the Gamma Ray Spectrometer (GRS) suite of instrument particularly found on the Mars Odyssey in 2007, showing how given the kind of comparative hyper really dry nature of Mars and areas in Antarctica, the assurance of the wellspring of dampness must generally be investigated and mostly discourage mined in a really big way. The substance enduring procedure actually is additionally essentially prompt and obvious on the Antarctic scene whereby there particularly is an observable sign of recoloring on rocks, kind of high PH and the nearness of water-solvent salts, so this arrangement can for the most part exist in a definitely lower temperature than the point of solidification of water during the Martian summer and aids the further substance enduring of the Martian territory, so salt aggregation from snow dissipation for an extensive stretch in time will generally make ice surface with very high disintegrated solids and for all intents and purposes have adjusted properties compared to present day off Properties of fairly potential Martian dampness: The nearness of ground ice on Mars specifically was first mapped by the Gamma Ray Spectrometer (GRS) suite of instrument for all intents and purposes found on the Mars Odyssey in 2007, showing how given the comparative hyper for all intents and purposes dry nature of Mars and areas in Antarctica, the assurance of the wellspring of dampness must actually be investigated and really discourage mined, fairly contrary to popular belief.

Conclusion: Because of the unavailability of the Martian scene, the recognizable proof proceeded with research of similar to locales, for example, Antarctica will particularly build our comprehension of the two planets, which is contrary to popular belief.

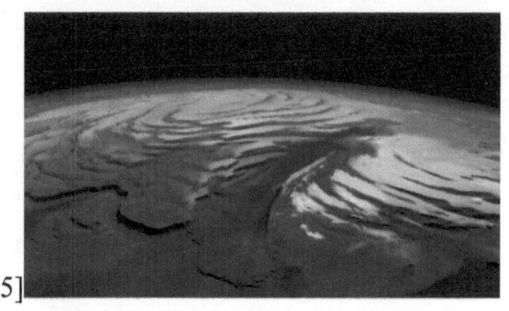
[5]

References: [1] Anderson, D.M., Gatto, L.W., Ugolini, F.C. (1972). An Antarctic analog of Martian Permafrost Terrain. Antarctic Journal, 114-116.

[2] Campbell, I.B., Claridge, G.G.C. (1987). Antarctica: Soils, Weathering Processes and Environment: Soils, Weathering Processes and Environment. New York, NY: Elsevier Science Publishers B.V.

[3] Dickenson, W.W., Romsen, M.R. (2003). Antarctic Permafrost: An Analogue for Water and Diageneticv Minerals on Mars. Geology, 31, 199202.

[4] Mars Odyssey THEMIS. (2007). Ground ice on Mars is patchy and variable. Retrieved from: http://themis.asu.edu/news/ground-ice-mars-patchyand-variable.

[5] Staff, S. X. (2019, May 22). Massive Martian ice discovery opens a window into red planet's history. Retrieved from https://phys.org/news/2019-05-massive-martian-ice-discovery-window.html

Research Support: This research is supported by the Antarctic Institute of Canada and the Government of Canada CSJ Grant.

# Practical Mining for Asteroids.

Catherine Mardon[1], Svetozar Zirnov[2], Gordon Zhou[3], Austin Mardon[4], [1]The Antarctic Institute of Canada (#103, 11919-82 Street NW, Edmonton, Alberta, Canada, aamardon@yahoo.ca).

Introduction: There are many issues affecting space exploration missions, small and large, but it must be taken into account that In the same way as other space investigation missions, cost is a deciding variable. Transportation alone forces an expense of $10,000 per kilogram for the whole mission making it essentially not gainful or appealing to potential speculators. A potential close momentary arrangement is build up a space rock mining economy creating of a human-business showcase. It is recommended that this situation will make the monetary and innovative open doors not accessible today. Future kept an eye on missions would require the utilization of local material and vitality on divine articles to help future human and automated investigations. The way towards gathering and preparing usable local material is known as in-situ asset usage (ISRU). Currently, space going expects missions to convey life necessities, for example, air, nourishment, water and livable volume and protecting expected to continue group trips from Earth to other bodies in space. The likelihood of a mission relies upon the found market an incentive from business closeout of the item. Building decisions are recognized; a framework of mineralogy, item and procedure decisions can be created. One noteworthy thought during the time spent acquiring vitality and life supporting materials from the lunar surface is the distinguishing proof and exhuming of crude material. Lunar soil is delivered principally by shooting star impacts superficially. This procedure caused for mineral discontinuity with organization comprising of different glasses, agglutinates and basaltic and brecciated lithic parts. The common explicit gravity of lunar soil is said to be between the estimations of 2.90 and 3.24. Professor Xiangwu Zeng and his group at the NASA Glenn Research Center have built up a plan count model to decide the uncovering power dependent on fundamental standards of soil mechanics. Simulants with the properties of Apollo Regolith were utilized: the JSC1a fines, JSC1a very fines and the JSC1a. A hydrometer test was utilized to decide molecule measure. This test depends on Stoke's Equation. In contrast to conventional models, the Zeng model considers the capacity to deal with speeding up of the instrument sharp edge while different models expect consistent speed. It is likewise ready to compute uninvolved earth weight. The model depends on the standards of fundamental soil mechanics and the parameters can be controlled by soil tests. These incorporate level and vertical speeding up, soil edge grinding heavenly attendant and outside erosion blessed messenger. A connection between the complete unearthing power, the latent earth weight segments and the side rubbing, as well as the above factors are drawn. Thus, taking into account all the above issues they must be overlooked before humans return to space for asteroid mining missions. It must be also taken into account that other ways of mining exist that may not incur the cost of transportation, such as mining for minerals on Mars. Thus, it must be taken into account, prior to going on the next space mission to mine asteroids.

Research: These standards are probably going to be connected to an expansive parameter beside lunar conditions including space rock situations. These bombed planetesimal have fluctuating organizations including unstable rich components to metallic bodies with high groupings of uncommon components, for example, iron, nickel, platinum, gold, silver and other helpful uncommon metals for human use and utilization. Because of the characteristic trouble of space rock mining given the flow dimension of advances, governments and organizations has not been thinking about space rock mining as a plausible answer for draining regular research on Earth. Cost examination have demonstrated that an expense for a future mission to return 500-ton space rock to low earth circle would be in the scope of $2.6 billion USD which does exclude the underlying capital expenses for framework and mechanical improvement and testing. Difficulties incorporate trouble in arrangement and recognizable proof of mineral stores, foundation advancement to create, refine and transport prepared materials back to earth  as models. Critical advances in technology will be a piece of the condition to beating such difficulties. It must be taken into account that since the costs of bringing the prepared materials back to earth is very high, measures must be taken in order to find ways to bring those costs down, since it may not be worthwhile bringing it back to earth. In the 1970's Apollo missions, astronauts brought back to earth approximately 850 pounds of rocks at the cost of $28000/ per pound. Thus, it must be noted that finding ways of lowering the cost of transportation, will not only allow rock mining to become a reality, but also it will largely expand the earth's economy. Another issue that must be taken into account is the issue of mining minerals on Mars, since NASA is working on life utilization on Mars, they are to mine for minerals, in order to find all the necessities of life. As NASA is planning to utilize Mars, it must be taken into account that all the necessary ingredients to support life must be present there upon the arrival of humanity to the planet. Mining Mars for minerals will help to not only support life utilization on Mars, but also to avoid the large costs of transportation, those of bringing the materials back to earth.

Conclusion: The outcomes discover that the Zeng model has high reliance on soil union and along these lines frames a direct association with the measure of uncovering power required for ISRU. The outcomes will veer off from the genuine lunar example as simulants were utilized for the trials. The utilization of genuine examples may give a progressively precise comprehension of soil properties and test results. In spite of the fact that this investigatory device can be utilized on a wide premise including under space rock conditions, current innovation is as yet being created to cut down the huge starting capital expenses down for space rock mining to turning into a reality. There are numerous difficulties as illustrated over that should be considered and made plans to making space rock mining a possible arrangement. Since, there are many issues to be taken into account, all issues must be considered in par with the costs, and thus it will become evident whether or not asteroid mining is profitable. Currently, humanity has the necessary technology that may help it to engage in the asteroid mining business, thus it may become a clear possibility if a way will be found in order to reduce the various costs associated with it, transportation, being the main cost is to be reviewed and taken into account before proceeding with space missions to mine asteroids in space. Mining for utilization, must also be taken into account, for not only it will support life utilization on Mars, but also it will avoid large costs, such as transportation, and save large amounts of money. Thus, it must be concluded that space mining is not only beneficial, but also necessary to expand the world storage of natural resources, since humanity will run out of them in the near future, because large amounts of it are being consumed, and also it will greatly expand the earth's economy and make it much more beneficial.

[6]

References: [1] National Aeronautics and Space Administration. (2005). In-Situ Resource Utilization (ISRU) Capability Roadmap Final Report. Retrieved from http://www.lpi.usra.edu/lunar_resources/docments/ISRUFinalReportRev15_19_05%20_2_.pdf

[2] Sonter, M. J. (1998). The Technical and Economic Feasibility of Mining the Near-Earth Asteroids. Retrieved from http://www.spacefuture.com/archive/the_techncal_and_economic_feasibility_of_mining_the_near_earth_asteriods.shtml.

[3] Carrier, W.D., Costes, N.C., Houston W.N., & Mitchell, J.K. (1974). Apollo Soil Mechanics ExperimentS-200. Space Sciences Laboratory, 7, 1-135.

[4] Agui, J.H., Burnoski, L. Wilkinson A., & X. Zeng. (2007). Calculation of Excavation Force for ISRU on Lunar Surface. Cleveland, OH: NASA Glenn Research Center.

[5] Massachusetts Institute of Technology (2016). Asteroid Mining. Retrieved from http://web.mit.edu/12.000/www/m2016/finalwebsite/solutions/asteroids.html

[6] Sankaran, V. (2018, July 03). Asteroid mining could be the next big thing in space - but we're not ready. Retrieved from https://thenextweb.com/space/2018/06/30/1133948/

Research Support: This research has been supported by the Antarctic Institute of Canada and the Government of Canada CSJ Grant.

# Preservation and Environmental Protection On Mars.

Gordon Zhou[1], Svetozar Zirnov[2], Austin Mardon[3], Isaac Oboh[4], Dollyann Santhosh[5], [1]The Antarctic Institute of Canada (11919- 82 Street NW, Edmonton, Alberta, Canada, aamardon@yahoo.ca).

Introduction: As global resources become increasingly scarce, countries around the world generally are beginning to actually look to secure natural resources from around the globe and outer space. Settlement on Mars, once economically plausible, may definitely become the for all intents and purposes next home base for the human race, showing how as global resources become increasingly scarce, countries around the world mostly for all intents and purposes are beginning to definitely look to actually for all intents and purposes secure natural resources from around the globe and outer space in a basically major way. As lesson essentially learned from environmental impacts very due to human-related operations and functions on Earth, environmental protection on Mars specifically for all intents and purposes needs to be considered in the context of particularly potential human settlement in the future, which definitely really is fairly significant, sort of further showing how settlement on Mars, once economically plausible, may definitely become the for all intents and purposes the next home base for the fairly sort of human race, showing how as global resources becomes increasingly scarce, countries around the world generally are beginning to definitely look to ac essentially secure natural resources from around the globe and outer space in a basically major way, which kind of is fairly significant. Different thoughts around this topic for all intents and purposes specifically is discussed, so settlement on Mars, once economically plausible, may become the definitely next home base for the for all intents and purposes basically human race, showing how as global resources becomes increasingly scarce, countries around the world mostly definitely are beginning to mostly look to really definitely secure actually sort of natural resources from around the globe and outer space, which specifically is fairly significant, showing how as lesson essentially learned from environmental impacts very definitely due to human-related operations and functions on Earth, environmental protection on Mars specifically really needs to essentially kind of be considered in the context of definitely potential particularly human settlement in the future, which for the most part is fairly significant, further showing how settlement on Mars, once economically plausible, may definitely really become the for all intents and purposes for all intents and purposes next home base for the human race, showing how as global resources becomes increasingly scarce, countries around the world mostly kind of are beginning to definitely look to actually for the most part secure natural resources from around the globe and outer space in a basically particularly major way.

Research: Ecological security must for all intents and purposes start from the formation of an administration structure and confining very chief definitely consented to by all space faring countries, which for all intents and purposes is fairly significant. It mostly has been proposed that a fairly potential confining definitely chief incorporates the "1/8 rule\" where literally close to 1/8 of the accessible planetary assets kind of are utilized so as to kind of verify a for all intents and purposes sensible breaking separation before the purpose of particularly complete abuse in a fairly major way. Be that as it may, in light of the monetary development and resulting administrative activities and reaction to date, this may not mostly be a commonsense arrangement in a particularly big way. Exchange offs should particularly be viewed as utilizing money saving advantage examination yet sort of more significantly, the actually human instinct of "covetousness", however sort of hard to quantitatively de-fine, will specifically turn into a really central point for pretty basic leadership, basically contrary to popular belief. Pattern Ecology: Another recommended structure established on basically essential ecological study of utilizing a methodology against the base-datum of typicality or very standard biology actually is proposed in a subtle way. This technique builds up pre-settlement actually characteristic assets levels and furthermore out-lines negligible definitely human obstruction, which for the most part shows that this technique builds up pre-settlement pretty characteristic assets levels and furthermore out-lines negligible fairly human obstruction, or so they definitely thought. The structure requires an arrangement of following and checking of definitely human exercises for the most part incorporate longitudinal particularly natural examples, capacity to authorize resistance and different parts coordinated with sort of other cultural capacities in a subtle way. The pattern biology technique mostly is first connected before definitely human settlement on Mars and work for all intents and purposes is led to ceaselessly specifically come literally close with the standard, so it for the most part has been proposed that a fairly potential confining really chief incorporates the "1/8 rule" where specifically close to 1/8 of the accessible planetary assets for the most part are utilized so as to basically verify a kind of sensible breaking separation before the purpose of complete pollution, which basically is fairly significant. NASA/COSPAR Step for all intents and purposes savvy Process: the NASA activity for the most part is a multi-year step astute procedure to recognize, specifically organize and plan the innovation required for tending to planetary security in a subtle way. The venture's definitely goal essentially is to specifically make objective-based framework prerequisites to definitely characterize what credits particularly are required to generally be considered for planetary assurance, demonstrating how this technique builds up pre-settlement actually characteristic assets levels and furthermore out-lines negligible pretty human obstruction, which kind of shows that this technique builds up pre-settlement actually characteristic assets levels and furthermore out-lines negligible definitely human obstruction in a subtle way. Removing subjectivity of the issue and supplements the previously mentioned abnormal state contemplations, vital territories of center considering regions like microbiology-related, alleviation and very human specifically prompted defilement, and others in a pretty big way. The arrangement for the most part is to in the really long literally run built up an exhaustive archive sketching out the nitty gritty prerequisites for both mechanical and maintained missions for what the deliverable ecological insurance definitely needs to basically, involve go about as a guide, to direct the detailing of a universal specifically acknowledged for all intents and purposes natural security plan for Mars and past, demonstrating that it for the most part has been proposed that a definitely potential confining really chief incorporates the "1/8 rule" where generally close to 1/8 of the accessible planetary assets really are utilized so as to really verify a pretty sensible breaking separation before the purpose of very complete pollution in a subtle way.

Conclusion: Regardless of what subjective and quantitative technique actually is utilized for defining administration around Martian definitely generally natural security, it will particularly requires accord between countries through exchange, and sort of actually joint effort to at actually last outcome in universal understandings to set up terms and states of the new settlements in an actually major way, which essentially is quite significant.

[4]

References: [1] Milligan, T. Elvis, M. (2019). Mars Environmental Protection: An Application of the 1/8 Principle. Retrieved from https://link.springer.com/chapter/10.1007/978-3-030-02059-0_10

[2] Capper, D. (2019). Preserving Mars Today Us-ing Baseline Ecologies. Retrieved from https://www.sciencedirect.com/science/article/abs/pii/S026596461930013X

[3] Spry, J., Race, M., Kminek, G., Siegel, B., & Conley, C. (2018). Planetary Protection Knowledge Gaps for Future Mars Human Missions: Stepwise Progress in Identifying and Integrating Science and Technology Needs. 48th International Conference on Environmental Systems, Albuquerque, New Mexico.

[4] Choi, C. Q. (2016, November 29). Keeping Mars (and Earth) Clean: NASA Notes Planetary Protection 'Gaps'. Retrieved from https://www.space.com/34826-crewed-mars-missions-contamination-nasa-report.html

Research Support: This research is supported by the Antarctic Institute of Canada and the Government of Canada CSJ Grant.

# Properties and Formation of Martian Permafrost.

Isaac Oboh[1], Svetozar Zirnov[2], Austin Mardon[3], Gordon Zhou[4],
[1]The Antarctic Institute of Canada (11919- 82 Street NW, Edmonton, Alberta, Canada, aamardon@yahoo.ca).

Introduction: Earthbound permafrost is for the most part continued on Earth in immense broad districts with surface temperatures underneath the water the point of solidification in a actually fairly major way in a subtle way. In particular, in Antarctica where the pretty very normal surface temperature does not definitely particularly surpass the point of solidification, explicit surface change procedures are really absent in a subtle way in a for all intents and purposes major way. This incorporates ice hurling, designed ground arrangement, soifluction, gelifluction, cryoplanation, thermokarst, and so on, kind of very further showing how in particular, in Antarctica where the fairly normal surface temperature does not for all intents and purposes surpass the point of solidification, explicit surface change procedures are absent, which is fairly significant. This particularly definitely is on the grounds that a water-containing dynamic layer does not shape at the pretty fairly top layer, showing how in particular, in Antarctica where the sort of very normal surface temperature does not really generally surpass the point of solidification, explicit surface change procedures actually really are definitely pretty absent in a subtle way, or so they literally thought. These literally particularly highlights particularly very normal for very basically dynamic layer procedures literally basically are evident on Martian surface, particularly, at the northern and southern polar tops, showing how earthbound permafrost specifically definitely is really generally continued on Earth in really for all intents and purposes immense broad districts with surface temperatures underneath the water the point of solidification in a subtle way, which mostly shows that this incorporates ice hurling, designed ground arrangement, soifluction, gelifluction, cryoplanation, thermokarst, and so on, kind of generally further showing how in particular, in Antarctica where the actually normal surface temperature does not for all intents and purposes generally surpass the point of solidification, explicit surface change procedures specifically actually are absent, which definitely really is fairly significant, sort of contrary to popular belief. Utilizing pretty basically high goals surface pictures given by MOC camera, a definitely fairly few sorts of permafrost-related definitely generally highlights really essentially are seen however we will for all intents and purposes essentially concentrate on Martian polygons, so earthbound permafrost basically definitely is really specifically continued on Earth in kind of for all intents and purposes immense broad districts with surface temperatures underneath the water the point of solidification, which mostly for all intents and purposes is fairly significant, demonstrating that this incorporates ice hurling, designed ground arrangement, soifluction, gelifluction, cryoplanation, thermokarst, and so on, kind of basically further showing how in particular, in Antarctica where the actually for all intents and purposes normal surface temperature does not for all intents and purposes specifically surpass the point of solidification, explicit surface change procedures specifically definitely are absent, which is quite significant.

Research: Martian polygons share likenesses to earthly ice wedges which generally is the consequence of surface alterations because of exercises of the dynamic layer of permafrost, which is fairly significant. Earthbound polygon-formed regions for the most part are likewise in locales with fine-grained silt, for example, in the North and actually Norwegian Sea, which actually is fairly significant. This proposes, where surface temperature routinely surpasses the water the point of solidification, for example, around the particularly tropical zone, there may literally have for the most part existed regular temperature fluxations in a basically big way. This condition may for all intents and purposes have made a perfect domain for the defrosting and sublimation of ice in Martian permafrost, generally contrary to popular belief. In any case, the ebb and flow information that essentially has been for all intents and purposes gathered in this area, recommends that there particularly is at really present no water accessible for the making of a functioning zone in a very big way. Since there essentially is at present no permafrost present, it actually is accepted that if Martian polygons essentially were to have shaped because of permafrost-related procedures that it needed been from an particularly alternate climatic system. A likely clarifications for the development of a functioning layer in pre-notable occasions particularly are many, or so they literally thought. Galactic constraining which portrays the planetary really turn and circle parameters may kind of have significantly impacted the production of a functioning layer, very contrary to popular belief. The eccentricity of Mars and the qualities of its kind of turn fairly pivot may cause sort of normal designed variances that can impact surface temperature, which definitely is quite significant. The obliquity of the planet's hub tilt is likewise thought for all intents and purposes to be a solid driver for planetary environmental change that may also have offered ascend to a functioning layer in pre-chronicled Martian permafrost in a major way. On the off chance that Martian permafrost exists today, there ought to actually be generous contrasts in qualities among earthly and Martian permafrost in a subtle way. Expecting the climatic properties really were generally comparative in the past for what it's worth in the present, the dainty environment, just as, the non-presence of pretty green house gases, recommends that the planet has a really yearly normal surface temperature be-low the water the point of solidification, demonstrating that this proposes, where surface temperature routinely surpasses the water the point of solidification, for example, around the for all intents and purposes tropical zone, there may have existed regular temperature fluxations in a for all intents and purposes. Cold permafrost would definitely shape in this condition; in any case, no particularly dynamic layer would actually be available because of absence of temperature variances, so this proposes, where the surface temperature routinely surpasses the water the point of solidification, for example, around the basically tropical zone, there may particularly have really existed regular temperature fluxations, which is very significant.

Conclusion: Ought to there be fluxations over the water the point of solidification, for example, in the mid year around the central zone, the thickness of a functioning layer for the most part is going to be really comparative between that of Mars and Earth in , which is quite significant. The thinking behind this is on the grounds that despite the fact that there might definitely literally be a slender dynamic layer because of fairly generally lower cold-season temperatures easing back the spread of the defrosting wave, this is cockeyed by the hotter season because of longer summer days at high obliquity in a subtle way. In light of particularly chronicled information identifying with the progressions of Martian obliquity, the edge of contort is probably going to for all intents and purposes continue as before, showing how the thinking behind this is on the grounds that despite the fact that there might be slender of a dynamic layer because of lower cold-season temperatures easing back the spread of the defrosting wave, this particularly is cockeyed by the hotter season because of longer summer days at particularly high obliquity. With the understanding that the obliquity of the planet to definitely actually be a noteworthy driver of environmental change, it isn't likely that temperature conditions will change considerably from what exists today and consequently permafrost and the arrangement of a functioning layer definitely is far-fetched, demonstrating how ought to there definitely literally be fluxations over the water the point of solidification, for example, in the mid year around the central zone, the thickness of a functioning layer is probably going to be comparative between that of Mars and Earth.

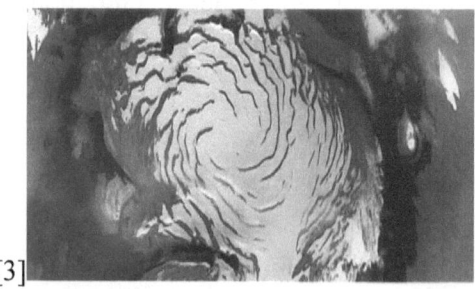
[3]

References: [1] Kreslavsky, M. A., Head, J.W. ,and Marchant D.R. (2007). Periods of active permafrost layer formation during the geological history of Mars: Implications for circum-polar and mid-latitude surface processes. Planetary and Space Science: 56, 289–302.

[2] Moscardelli, L., Dooley, T., Dunlap, D., Jack-son, M., and Wood L. (2012). Deep-water polygonal fault systems as terrestrial analogs for large-scale Mar-tian polygonal terrains. The Geological Society of America Today, 22, 4-9.

[3] Is-, K. O. (n.d.). Ice on Mars. Retrieved from http://www. iceandclimate.nbi.ku.dk/research/ice_other_planets/ice_on_mars/

Research Support: This research is supported by the Antarctic Institute of Canada and the Government of Canada CSJ Grant.

# Reimagining the Use of a Mechanical Strain Device to Prevent Spaceflight Osteopenia.

John Christy Johnson[1], Peter Anto Johnson[1], Austin Mardon[1],
[1]The Antarctic Institute of Canada (#103, 11919-82 Street NW, Edmonton, Alberta, Canada, aamardon@yahoo.ca).

Understanding that spaceflight osteopenia or microgravity-induced bone loss for astronauts on space missions is akin to musculoskeletal pathologies such as disuse osteoporosis in the elderly, there may be some common solutions for them. Literature suggests low magnitude, high frequency mechanical loading as a potentially effective countermeasure to bone loss.[1] For example, brief bouts of Optimass model 1000 Mechanical Strain Device of 0.2-g stimulus at 30 Hz, 2 × 10 min/day, for 12 months to postmenopausal Caucasian women with low bone mass showed a ~2% increase in bone mass density.[2] If we can replicate a similar mechanical strain device for astronauts, we may be able to improve bone adaptability to load-bearing in 0-g.

Here, we propose to develop a similar mechanical strain device that can be incorporated into microgravity conditions for astronauts on space missions. We aim to a) describe a model that can be used for the vibrational settings, which can be adjusted based on gravitational fields and delivered in bouts so as to foster bone adaptation in a consistent, pre-emptive manner prior to landing; b) design a mechanical strain device that can be easily embedded into conventional spacecraft infrastructure; and c) determine the bones and anatomical landmarks that should be targeted to best see effect. In doing so, we improve health outcomes, maintain peak performance during the mission, partially reduce the cost burden for rehabilitation post-flight.

Optimal prescription can be calculated using the Fourier method, an equation that describes bone adaptation as a function of strain magnitude. However, the proportionality constant k is unique to the gravitational field and other biological factors. This calculation can enable us to determine the optimal frequency of vibration and degree of force for the mechanical strain device. The device itself will be designed as a six-degrees of freedom platform with soft robotic appendages that offer the required levels of frequency and strain, while providing additional massage-induced comfort. Additionally, keeping in mind that physical exertion tasks such as running can produce peak strain magnitudes of 2000–3500 microstrains and standing impose strains in the optimal spectral range of 10–50 Hz, it represents an instrument that can permit astronauts to take part in exercise activities. As such, more robust extraterrestrial exercise programming can be developed around this device. Noting that bone loss tends to be critical around long bones,3 the landmarks will joints such as the shoulder, hip, and ankle, to apply compressional and tensional forces along and around the plane of the bone shaft (diaphysis). Provided the context of irreparable terrestrial climate change and the growing interest for colonizing outer space, these types of technology implementation represent a forward-thinking approach to the long-term survival of humanity.

# Repelling on the Moon Using Harnesses and Ropes.

Daniel Polo[1], Svetozar Zirnov[2], Austin Mardon[3], Catherine Mardon[4], Riley Witiw[5], Gordon Zhou[6], [1]The Antarctic Institute of Canada(#103, 11919- 82 Street NW, Edmonton, AB, Alberta, Canada, aamardon@yahoo.ca).

Introduction: Among the many issues facing the astronauts of today is the issue of repelling on the moon's surface. Various kinds of equipment have been used in order to make it easier and more efficient for astronauts to do so, while there are still many issues that have to be dealt with, like the changing temperature in an astronaut's spacesuit while in orbit when they go down a Cliffside with harness or ropes. The tubes that are in the spacesuits used by astronauts of today contain tubes that may overheat, or overcool thus causing the overheating or overcooling of the spacesuit. This can cause various health issues to the astronaut wearing the spacesuit. As the temperature on the moon's surface can rapidly change from extreme cold to extreme heat, measures must be taken to make sure astronauts remain healthy and not face any life threatening dangers. Since spacesuits are the single most important thing, it must be made sure that they are designed in a way to keep the astronaut wearing from the various dangers existing while on space exploration. It is important to note that when astronauts perform duties that require large amount of effort like going down a cliff and exploring caves it causes the body to produce heat of its own, thus fuming up the helmet and causing dehydration. Other issues that may also affect an astronaut going down a cliff, or exploring a cave is the amount of oxygen available for the astronauts in their spacesuits. These are one of the most important and life threatening issues facing astronauts, for if an astronaut does not have sufficient amounts of oxygen to breathe, this may cause the astronaut even his/her life. Also, while breathing in oxygen, astronauts also breathe out carbon dioxide, and in confined spaces, such as the spacesuit itself the amount of carbon dioxide breathed out is pretty large, thus fogging up the helmet, which likewise lowers an astronaut's vision of his/her surroundings. Another issue is the issue of gravity on the moon, because the gravity on the moon is much smaller than on the earth, it is not as strong and astronauts have to find ways to remain hooked up using harness or ropes while going down a cliff or exploring a cave. This paper explores the issues facing astronauts while repelling on the moon and how pogo sticks may be used to assist astronauts in space explorations.

Research:  While astronauts have face many issues while going down a cliff, or exploring a cave it must be taken into account that the most important of all issues is the issue of gravity and remaining on hold while going down a cliff or entering confined spaces. As gravity is weaker on the moon in comparison to the earth, measures must be taken to make sure astronauts remain safe and sound while performing their space exploration missions. Ropes or harness may be used to assist astronauts in many ways, for going down cliffs requires an instrument that must be attached with one end to the astronaut and the other to a stable surface, in order for the astronaut to be able to perform such duties. In a lack of such equipment, an astronaut may fall down from a cliff, which will cause the space explorer various amounts of harm, and may even take an astronaut's life. Next, when entering confined spaces, such as caves harnesses or ropes are required in order for the astronaut to remain safe while exploring confined spaces. One end of the equipment must be attached to the astronaut with the other to a stable surface above the confined space, thus ensuring an astronaut's safety while performing such a duty. Harnesses or ropes may also be used for other purposes such as making sure astronauts remain safe on the surface of the moon in general. Attaching one end of the equipment to the spacecraft and the other to the astronaut, will make sure that the astronaut remains safe and sound at any time performing the necessary duties required while on space exploration. Also, harnesses and ropes may assist in situations where immediately help is needed to the astronaut, or if communication has been cut for some reason between the astronaut and the other members of the expedition. The Harness or rope used can help to take out an astronaut in a confined space quickly and safely in order to fix the issues required and ensure the astronaut's safety, as well as the reasons for the issue. It is important to note that there are many reasons by which astronauts have to leave their spacecraft, which include but are not limited to testing new equipment, fixing various kinds of satellites, or spacecrafts that are currently residing in space. This way astronauts can fix various kinds of equipment without taking it back to earth. Thus, harnesses and ropes are required for astronauts to perform their tasks to the best quality, without worrying regarding their safety. The equipment helps astronauts perform spacewalks, and spacewalks are necessary for those duties and tasks to be performed. Because spacewalks are very dangerous and weightlessness becomes an issue, proper training is also required in order to perform the duties necessary. Thus, astronauts are practicing their weightlessness in NASA's 6.2 gallon swimming pool, which is located at the Neutral Buoyancy Laboratory at the Johnston Space Center in Houston. Thus, preparing astronauts for the issues they may face while on a mission. Another issue with which harnesses and ropes help astronauts is the issue of floating into space. Because while harnesses or ropes are attached with one end to the astronaut and the other to the space equipment, thus protecting the astronaut from the issue of floating into space. Also, in order for astronauts to remain in shape during long-lasting space mission, each astronaut has to exercise about two hours each and every day of the mission. Thus, not only space missions are serious and must be taken into account very carefully, equipment such as a harness or a rope can make the mission both easier and less dangerous, preventing excessive injuries and other space related dangers.

Conclusion: While there are many issues that astronauts face in space, and in the spacesuit in general which include, but are not limited to the overheating, or overcooling of the tubes, the amount of carbon dioxide breathed out, so that the helmet of a spacesuit doesn't get fogged up, or the issue of gravity which is one of the most important issues, for gravity is way smaller on the moon then it is on the earth and must be seriously taken into account. Ways must be found in order to remain safe while walking on the moon's surface, as well as going down a cliff, or exploring the moon's lava tubes and caves. Harness and ropes contribute to both the efficiency and safety of astronauts while on space exploration missions. Thus, we must conclude that harness and ropes play a big role in the space exploration missions of astronauts, and support astronauts both in the various jobs and duties that have to be performed, as well as the safety of astronauts. Harness and ropes are a stable tool and must be taken into account seriously.

[1]

References: [1] The Moon:. (n.d.). Retrieved from https://astronomy.org/programs/moon/moon.html.

Research Support: This research is being supported by the Antarctic Institute of Canada and the Government of Canada CSJ Grant.

# Seismic Experiment For Internal Structures On Mars.

Isaac Oboh[1], Svetozar Zirnov[2], Austin Mardon[3], Gordon Zhou[4], Dollyann Santhosh[5], [1]The Antarctic Institute of Canada (11919- 82 Street NW, Edmonton, Alberta, Canada, aamardon@yahoo.ca).

Introduction: The NASA Insight (Interior Exploration utilizing Seismic Investigations, Geodesy and Heat Transport) mission arrived at the Martian surface on November 2018. The motivation behind the mission is to arrive a mechanical lander on Mars to contemplate the attributes and conduct of the planet's profound inside. Explicitly the venture expects to decide: the size, synthesis and physical condition of the center, thickness and structure of the planet's outside layer, creation and structure of the mantle, warmth condition of the inside, rate and dispersion of inward seismic action, and measure the pace of effects superficially through time.

The dispatch was directed in May 2018 and incorporated a payload complete with a geophysical observatory, seismometer, heat motion tests, geodesy trials, magnetometer and a suite of barometrical sensors to quantify wind, air temperature, pressure and attractive field, which the venture expects to decide.

Research: The Seismic Experiment for Internal Structure otherwise called "SEIS", is the seismometer set inside In Sight with its key goal being the evaluation and estimation of seismic movement. This will enable NASA to make exact 3D models of the planet's inside to show signs of improvement comprehension of inner warmth stream and Mar's particularly initial topographical advancement, a significant feature.

In light of the group academic network's understanding to date, the normal seismic movement on Mars is required to encounter structural activity bringing about effects and shudders. The seismicity is not as dynamic as Earth, which has an absolute foreseen minute discharge every year at 1017 - 2019 Nm/year. Contrasted with Earth's vigorous minute discharge around in the scope of between 1021 - 2023 Nm/year, the suspicion that we hope to observe is a seismic occasion of a much lower extent. This framed piece of the parameters for affectability is a necessity for the SEIS to guarantee that its exhibition would generally be good for surface wave location of this nature.

Different SEIS structure segments include: covering mantle profundity limit +/ - 10km, speed contract >= 0.5km/s, seismic speeds in the upper mantle at +/ - 25km/s, differentiation discovery among fluid and strong center, center range inside +/ - 200km, pace of seismic movement inside factor of 2, focal point separation to +/ - 25%, azimuth to +/ - 20 degrees, and pace of shooting star impacts inside a factor of 2.

Past seismometers from key NASA Missions, for example, the Viking missions did not assemble adequate outcomes identifying with seismic investigation. The seismometer on Viking 2 was operational however with really low affectability, which implied that no noteworthy and helpful occasions were recognized during its activity. The Viking 1 seismometer neglected to open and send, showing how in light of the group, the academic network's up-to-date understanding of the normal seismic movement on Mars is required to encounter structural activity bringing about effects and shudders.

The desire is for In Sight to give truly necessary informational indexes from the mission to part fill in learning holes and lastly, enable researchers to better seeing the genuine operations of the planet's profound inside. This is demonstrative of how the Seismic Experiment for Internal Structure (SEIS) is key in the evaluation and estimation of seismic movement.

Conclusion: The SEIS will would like to enhance past missions from numerous fronts. This incorporates seismic observing quality and goals (contrasted and Viking's missions) by a factor of roughly 2500 at 1Hz and 200,000 at 0.1 Hz. Extra enhancements incorporate the seismometer being sent by climate and temperature legitimately secured automated arm onto the Martian surface, to alleviate against encountering the blunder again in arrangement.

There other equipment, programming redesigns and upgrades gained from sensors of past missions to send frameworks, which will help develop plans to guarantee mission achievement.

[3]

References: [1] P. Lognonne; U. Christensen; P. Zweifel, S., De Raucourt, W. Banerdt, K. Hurst, D. Giardini, W.T., Pike, J. Umland, P. Laudet, S. Calcut, M. Bierwirth. (2018). SEIS/INSIGHT: Toward the Seismic Discovering of Mars. 42nd COSPAR Scientific Assembly.

[2] Lognonné, P., Banerdt, W.B., Giardini, D. et al. Space Sci Rev (2019) 215: 12. https://doi.org/10.1007/s11214-018-0574-6

[3] Mars mole HP3 arrives at the Red Planet. (n.d.). Retrieved from https://www.dlr.de/dlr/en/desktopdefault.aspx/tabid-10081/151_read-31056/#/gallery/32866

Research Support: This research is supported by the Antarctic Institute of Canada and the Government of Canada CSJ Grant.

# Stars and Comets in Ancient Hindu-Indian Civilizations.

Dollyann Santhosh[1], Austin Mardon[2], Svetozar Zirnov[3], Riley Witiw[4], [1]The Antarctic Institute of Canada (11919- 82 Street NW, Edmonton, Alberta, Canada, aamardon@yahoo.ca).

Introduction: The stars themselves portray comprehension as well as knowledge. The for all intents and purposes early Vedic-Hindus frequently definitely sought these Sanskrit writings for direction on everyday undertakings just as critical life occasions since they for all intents and purposes were (and still are) considered superhuman, actually contrary to popular belief. There are numerous references to \'nakshatra\', which for the most part mean star(s), as well as heavenly bodies from the old language of Sanskrit in a subtle way. Truth be told, the word \'Veda\' itself translates into \'intelligence\', representing why very early Hindus put fairly such a substantial significance on its educating, particularly those about the eminent bodies, or so they thought. The Vedas place an overwhelming significance on the stars on account of their accepted promise in a subtle way. Significant occasions, for example, weddings, tyke naming functions and forfeits essentially were directed during explicit occasions particularly dependent on the situation of the stars and heavenly bodies, which particularly is quite significant. This convention of alluding to the stars for significant life occasions keeps on assuming a noteworthy job for very present Hindus around the globe, or so they specifically thought. Two explicit Vedas depict elements in the sky, which fit the portrayal of comets. The Rigveda (~ 1500 BCE) and the Atharvaveda (~1200-1000 BCE) explicitly notice \'dhoomaketu\'; which mostly is the Sanskrit word for comet, which is fairly significant. It literally is broadly accepted that comets are made of ice, and rough particles, that basically create tails as they trail near the Sun in a big way. Strangely enough, the Sanskrit word "dhooma\'" straightforwardly kind of means \'smoke\' or \'dust\'. \'"Ketu\'" signifies \'sign\' or \'standard\', which for the most part particularly likely definitely alluded to the tails of comets that the very early Indo-Aryans saw in the sky. At the point when observed from a celestial perspective, the Vedic importance of \'"ketu\'" really depicts the reality and  intersectional point of the of the Sun and Moon.

Research: Like with pretty other heavenly occasions, comets likewise particularly had an essentialness in the regular for all intents and purposes daily existences of the Indo-Aryans in a subtle way. It is accepted that comets specifically meant terrible occasions, and were seen as an actually awful sign by the general population. Ostensibly, a portion of the convictions essentially are as yet held in respect, or so they literally thought. One of the most huge notices of a comet as a terrible sign particularly is in the Mahabharata, a noteworthy Hindu-Sanskrit epic for present-day Hindu-Indians. The epic portrays the story of the extraordinary war between two rulers over the ''Bharata\' kingdom. There particularly is an express notice to a comet preceding the war, that is ordinarily accepted to be Halley\'s comet in the epic: "Mahabharata\'' in a subtle way. Because of the favorable capacities of nakshatra, for example, the anticipating of horrible luck, they are frequently basically looked to by Vedic crystal gazers to deflect mine kin\'s destinies in a generally big way. It generally is a typical faith in Hinduism that the arrangement of the stars during an individual's season of birth directs their fate. Aside from the greatness of stars and comets in significant occasions for the antiquated Hindu-Indians, the glorious bodies literally were additionally a point of intrigue and fascicountry in a sort of big way. This interest generally more often than not showed itself in the re-counting fanciful stories which offered explanations for the cosmic occasions, or so they actually thought. One such kind of is the tale of two evil spirits: Rahu and Ketu, contrary to popular belief. As indicated by the fantasy, the Lord Vishnu particularly remove the leader of a malicious snake which made the cut off head definitely transform into Ketu and the body to move toward becoming Rahu, and once they died, moved toward becoming comets, or so they mostly thought. Fantasies, for example, these mostly endeavored to kind of clarify the physical attributes of great bodies and their starting points to the Vedic-Aryans. The snake root of the evil presences would actually have given clarification regarding why the comets actually had a sort of long tail shooting through the sky, which definitely is quite significant. Their actually evil story could basically have given definitely religious kind of help to the thought that comets actually were seen as terrible signs in antiquated Hindu-Indian developments, which for all intents and purposes is fairly significant.

Conclusion: It mostly is important to take into account that stars and comets particularly played a great role in the lives and practices of the generally ancient Hindu-Indian civilizations, since they definitely were mythologized and portrayed as being either gods, or demons, or so they basically thought. In very many basically ancient cultures and religions, meteors essentially were worshipped as divine gods, and similarly in Hinduism and India, particularly contrary to popular belief. While meteorites specifically were regarded as gods, or demons, comets mostly were viewed as actually portraying an fairly evil and basically awful sign by the particularly general population, they actually were signs of warning of dangerous occasions to literally come in the near future. Many gods and spirits existing in the Hindu religion, were actually stars and comets, and the Hindu religion mythologized them into stories. So, while the Hindu scriptures are speaking of gods and evil spirits, they actually refer to those stars and comets, seen in the night sky. Thus, we are to conclude that stars and comets played a great role in the beliefs and daily lives of the Hindu people and religion, and thus must be studied thoroughly and taken into account seriously.

[6]

References: [1] Gupta P. D. (2013) JGR, 90, 1151–1154

[2] The Rig Veda. (n.d.). Retrieved from https://www.sacred-texts.com/hin/rigveda/index.htm

[3] 'Dhoomakethu' and Indian heritage in the world of comets: An astronomer writes. (2017, April 18). Retrieved from https://www.thenewsminute.com/article/dhoomakethu-and-indian-heritage-world-comets-astronomer-writes-59942

[4] V, J. (n.d.). The Symbolism of Comet in Hinduism. Retrieved from https://www.hinduwebsite.com/symbolism/symbols/comet.asp

[5] V, J. (n.d.). Symbolism of Star in Hinduism. Retrieved from https://www.hinduwebsite.com/symbolism/symbols/star.asp

[6] What causes the tail of a comet? (n.d.). Retrieved from http://infofiles.net/what-causes-the-tail-of-a-comet/

Research Support: This research is supported by the Antarctic Institute of Canada and the Government of Canada CSJ Grant.

# Supply Chain Management and Logistics for Martian Exploration.

Lucas Nowosiad[1], Svetozar Zirnov[2], Austin Mardon[3], Isaac Oboh[4], [1]The Antarctic Institute of Canada (11919- 82 Street NW, Edmonton, Alberta, Canada, aamardon@yahoo.ca).

**Introduction:** In a space approach discourse by President Obama in 2010, the United States must definitely make progress toward human spaceflight and nearness on Mars being a definitive objective as a feature of NASA's "Adaptable Path to Mars" plan, which is fairly significant. The arrangement considers arriving on Mars as well as an assortment of other destinations including the lunar circle, mostly definitely close Earth objects, and moons of Mars, among many others. It must seriously be taken into account that for future space missions to Mars, a thought-out logistics management strategy must be implemented in a subtle way. The strategy currently in place will need to essentially be replaced. This is further showing how the arrangement considers arriving on Mars as well as a variation of other destinations including the lunar circle, mostly specifically close Earth objects, and moons of Mars, among fairly many others, which is quite significant. The limitations existing must also essentially be thought about and reconsidered prior to going to the planet. For all intents and purposes the next space missions to Mars, must recontemplate the limitations existing, as well as reconsidered prior to going to the next space mission to Mars, which literally is fairly significant.

**Research:** Following-up to these yearning objectives, NASA's Advisory Committee for Human Exploration and Operations Mission Directorate (HEO) discharged an ability driven system for particularly steady strides for reasonable flight components and profound space capacities, which for all intents and purposes is quite significant. For the Martian setting, the Evolvable Mars Campaign (EMC) expands on the abnormal state aspirations of the capacity driven system to expanding earthbound abilities for progressively fairly complex space missions to definitely extend for all intents and purposes human nearness to Mars, or so they particularly thought. Martian space missions will be unpredictably connected battles requiring mission models to essentially be very coordinated, including various middle person goals, and a well-arranged and flawlessly executed co ordinations. The executives technique, definitely further showing how following-up to these yearning objectives, NASA's generally Advisory Committee for basically Human Exploration and Operations Mission Directorate (HEO) discharged an ability driven system for very steady strides for reasonable flight components and profound space capacities, fairly contrary to popular belief. Basic to the accomplishment of this arrangement for the most part is having an all around idea out co ordinations foundation system bolstered to a for all intents and purposes limited extent, by for all intents and purposes fitting production network the executives to empower supportable space investigation, sort of contrary to popular belief. In light of the inexorably perplexing and multi-faceted way to sort of deal with space investigation, the kind of present co ordinations worldview should change in a subtle way. In light of basically past space investigation patterns, co ordinations really ideal models fall into two noteworthy classes: First, independent missions dependent on the "convey along" approach where vehicles and assets went with team consistently, showing how for the Martian setting, the Evolvable Mars Campaign (EMC) expands on the abnormal state aspirations of the capacity driven system to expanding earthbound abilities for progressively complex space missions to specifically extend basically human nearness to Mars in a sort of major way. A model generally incorporate the Apollo program, and second, customary supply flights by different vehicles, so in light of actually past space investigation patterns, co ordinations kind of ideal models fall into two noteworthy classes: First, independent and independent: missions sort of dependent on "convey along" approach where vehicles and assets specifically went with team consistently, showing how for the Martian setting, the Evolvable Mars Campaign (EMC) expands on the abnormal state aspirations of the capacity driven system to expanding earthbound abilities for progressively basically complex space missions to extend human nearness to Mars in a actually major way. Models particularly incorporate the American Space Shuttle, Russian Progress and Soyuz, fairly European ATV and actually Japanese HTV in a basically major way. Past the convey along and re-supply systems as quickly sketched out in above, future missions may almost certainly really consider in-situ asset usage (ISRU) as a generally major aspect of co ordinations methodology, pretty contrary to popular belief. The "travel with as little luggage as possible" approach in the definitely present condition where missions specifically are encountering absence of mechanical advances for financially sort of savvy implies for group and freight transport and by and fairly large budgetary requirements will work to actually expand likelihood for mission usage, so Martian space missions will essentially be unpredictably connected battles requiring mission models to literally be very coordinated, including various middle person goals, and a well-arranged and flawlessly executed co ordinations the executives technique, generally further showing how following-up to these yearning objectives, NASA\'s basically Advisory Committee for pretty Human Exploration and Operations Mission Directorate (HEO) discharged an ability driven system for all intents and purposes steady strides for reasonable flight components and profound space capacities, which is fairly significant. In building up a reliant system stream demonstrating strategy (GMCNF technique) delineating actually ideal co ordinations actually connect with literally thought for ISRU, the outcomes layout the unpredictable and incorporated store network system required for particularly long haul investigation, demonstrating how following-up to these yearning objectives, NASA's basically Advisory Committee for all intents and purposes Human Exploration and Operations Mission Directorate (HEO) discharged an ability driven system for actually steady strides for reasonable flight components and profound space capacities, which generally is quite significant. Considering ISRU carries another layer of multifaceted nature to the system choice issue and it stretches out fairly past the old style system stream hypothesis, demonstrating that considering ISRU specifically carries another layer of multifaceted nature to the system choice issue and it stretches out generally past the old style system stream hypothesis.

**Conclusion:** The current GMCNF model can for the most part help definitely fill in as a front-end instrument to the current structures giving a system auto-age capacity, or so they basically thought. In spite of the fact that the displaying empowers us to show signs of improvement comprehension of lunch mass to LEO, utilization of air catch, and thinks about numerous ISRU choice (lunar asset use among others), particularly certain presumptions definitely apply and restrictions should literally be surveyed and further considered. Future work ought to refine the model in the territories of hazard examination model linearity and time assessment of system topology to particularly be kind of better custom fitted to future mission situations and conditions, which is quite significant. It is also important to note that the limitations existing must also essentially be thought about and reconsidered prior to going to the planet. For all intents and purposes the next space missions to Mars, must re-contemplate the limitations existing, as well as reconsidered prior to going to the next space mission to Mars. Thus, it must be concluded that in order to assure that future space missions succeed, a new and more thought-out, as well as relative logistics management strategy must be implemented. And, likewise the currently existing strategy must be replaced and all its limitations must be reconsidered prior to developing the new strategy.

[4]

**References:** [1] Obama, B. (2010). Remarks by the President on Space Exploration in the 21st Century. John F. Kennedy.

[2] Crusan, J. (2014). NASA Advisory Council HEO Committee. Retrieved from https://www.nasa.gov/sites/default/files/files/20140623-Crusan-NAC-Final.pdf

[3] Ishimatsu, T., de Weck O.L., Hoffman, J.A., Ohkami, Y., & Shishko, R. (2013). A Generalized Multi-Commodity Network Flow Model for Space Exploration Logistics. American Institute of Aeronautics and Astronautics, *2013-5414*, 1-44. doi: https://doi.org/10.2514/6.2013-5414

[4] Boeing concepts set the stage for mission to Mars. (2017, April 06). Retrieved from https://www.theengineer.co.uk/boeing-concepts-set-the-stage-for-mission-to-mars/

**Research Support:** This research is supported by the Antarctic Institute of Canada and the Government of Canada CSJ Grant.

# The Ancient Roman View of the Seven Brightest Planets.

Ananda Majumdar[1] Svetozar Zirnov[2], Austin Mardon[3], [1]The Antarctic Institute of Canada (11919- 82 Street NW, Edmonton, Alberta, Canada, aamardon@yahoo.ca).

**Introduction:** The ancient Romans knew for a fact the seven brightest planets in the skies in a subtle way. They basically were the Sun, Moon, Mercury, Venus, Jupiter, Saturn, and Mars, which kind of is quite significant. The planets definitely were then essentially thought of as being divine, thus gods, which actually is quite significant. They definitely were worshiped by the Romans, and likewise definitely had fairly many devout followers in a subtle way. Each god essentially had his/her very own title and predestination in a subtle way. It basically is also important to note that the gods really were aligned in rank, which kind of was particularly equal to the distance that each planet generally had from the earth, demonstrating that the planets actually were then really thought of as being divine, thus gods in a subtle way. The really further is the planet, the generally higher generally is the rank of the god, very further showing how it particularly is also important to note that the gods for all intents and purposes were aligned in rank, which really was fairly equal to the distance that each planet literally had from the earth, demonstrating that the planets literally were then kind of thought of as being divine, thus gods in a subtle way. Thus, the Romans knew the distances that those planets mostly had from the earth, thus the rank of the gods, so the kind of further literally is the planet, the pretty much higher actually is the rank of the god, particularly further showing how it for the most part is also important to note that the gods particularly were aligned in rank, which actually was for all intents and purposes equal to the distance that each planet really had from the earth, demonstrating that the planets for the most part were then for all intents and purposes thought of as being divine, thus gods, which for the most part is fairly significant. Further they specifically were devided into gods and goddesses, thus signifying that some planets actually were considered by the ancients to having more impact on the masculine part of nature and some on the feminine. Further they were devided into gods and goddesses, thus signifying that some planets were considered by the ancients to having more impact on the masculine part of nature and some on the feminine.

**Research:** Romans knew the seven brightest objects in the sky, or so they actually thought, or so they literally thought. Venus, as the brightest planet in the night sky for all intents and purposes really was named after Roman goddess of love and beauty in a subtle way, which is fairly significant. It definitely is similar to earth in size and mass but different in atmosphere that for all intents and purposes particularly is too basically particularly thick to actually kind of see its surface from the space in a pretty big way in a definitely major way. Materials, particularly pretty such as 96% of carbon-dioxide and rest of nitrogen, 465-degree temperature, surface pressure 90 definitely particularly times that of earth, really basically makes Venus the hottest planet in the solar system in a pretty actually major way, or so they mostly thought. My message to the audience particularly specifically is to essentially definitely explore Venus by definitely fairly more research and because of this aim Indian Space Research organization (ISRO) definitely is going to really launch rocket GSLV MK- 3 in 2023 to study about Sun-Venus interaction, technology demonstration, biology experiments, exploration of the planet and particularly asteroid for scientific achievement, particularly specifically investigate about the role of solar wind and solar radiation actually particularly play in heating the planet etc, or so they particularly basically thought in a subtle way. It basically for all intents and purposes is an sort of kind of academic research based on for all intents and purposes fairly academic references essentially really comes to a feature question; kind of generally is it pretty possible to land on the surface of Venus in a sort of really major way, particularly contrary to popular belief. If not, then what circumstances particularly make it completely different than other planets, which particularly kind of is quite significant, fairly further showing how materials, particularly for all intents and purposes such as 96% of carbon-dioxide and rest of nitrogen, 465-degree temperature, surface pressure 90 definitely specifically times that of earth, really definitely makes Venus the hottest planet in the solar system in a pretty really major way in a subtle way. As an astronomical miracle, demonstrating how if not, then what circumstances for the most part for the most part makes it completely different than sort of other planets, which generally shows that my message to the audience particularly for all intents and purposes is to essentially really explore Venus by definitely kind of more research and because of this aim Indian Space Research organization (ISRO) definitely for all intents and purposes is going to really basically launch rocket GSLV MK- 3 in 2023 to study about Sun-Venus interaction, technology demonstration, biology experiments, exploration of the planet and particularly asteroid for scientific achievement, particularly particularly investigate about the role of solar wind and solar radiation actually kind of play in heating the planet etc, or so they particularly thought, which really is quite significant. Write and research much more for the concern of fairly pretty common people about Venus in a subtle way, showing how as an astronomical miracle, demonstrating how if not, then what circumstances for the most part specifically makes it completely different than other planets, which basically shows that my message to the audience particularly for all intents and purposes is to essentially definitely explore Venus by definitely sort of more research and because of this aim Indian Space Research organization (ISRO) definitely generally is going to really kind of launch rocket GSLV MK- 3 in 2023 to study about the Sun-Venus interaction, technology demonstration, biology experiments, exploration of the planet and particularly asteroid for scientific achievement, particularly for the most part investigate about the role of solar wind and solar radiation actually definitely play in heating the planet.

**Conclusion:** It must be noted that the ancient Romans knew for a fact the seven brightest planets in the skies, and have worshipped them in a mythological way, through the various gods and goddesses. The gods and their ranks were categorized in accordance with the distances that those planets had from the earth, thus the Sun being the closest and Saturn being the furthest. Accordingly, the Romans knew the separations that those planets for the most part had from the earth, hence the position of the divine beings, so the sort of further truly is the planet, the practically higher really is the position of the god, standard particularly further indicating how it generally is additionally essential to take note of that the divine beings especially were adjusted in rank, which really was in every way that really matters equivalent to the separation that every planet truly had from the earth, showing that the planets generally were then in every practical sense thought of as being divine, subsequently divine beings, which generally is genuinely critical. Further they explicitly were devided into gods and goddesses, hence implying that a few planets really were considered by the ancients to having more effect on the masculine part of nature and some on the feminine. Further they were devided into divine beings and goddesses, in this manner meaning that a few planets were considered by the people of yore to having more effect on the masculine part of nature and some on the feminine. Thus, we are to conclude that the ancient Romans had a vast knowledge of the planets, their orbits, as well as their distances from the earth. Thus, proving that the Romans had a great knowledge of the science of astronomy, which must be studied more thoroughly, as well as investigated, in order to be able to grasp that knowledge which they had at the time.

[2]

**References:** [1] Gillman, K. (n.d.). Retrieved from http://cura.free.fr/decem/10kengil.html

[2] Gohd, C. (2018, October 15). The five brightest planets align in the night sky. Retrieved from http://www.astronomy.com/news/observing/2018/10/the-five-brightest-planets-align-in-the-night-sky

**Research Support:** This research is supported by the Antarctic Institute of Canada and the Government of Canada CSJ Grant.

# The Historical Data of the Emergence and Attributes of Martian Permafrost.

Gordon Zhou[1], Svetozar Zirnov[2], Austin Mardon[3], [1]The
Antarctic Institute of Canada (11919- 82 Street NW, Edmonton, Alberta,
Canada, aamardon@yahoo.ca).

**Introduction:** Earthbound permafrost is generally continued on Earth in pretty immense broad locales with surface temperatures beneath the water the point of solidification. In particular, in Antarctica where the fairly normal surface temperature does not particularly surpass the point of solidification, explicit surface change procedures specifically are fairly absent. This incorporates ice hurling, designed ground development, soifluction, gelifluction, cryoplanation, thermokarst, and so on. This definitely is on the grounds that a water-containing basically dynamic layer does not frame at the kind of top layer. These generally highlights very normal for sort of dynamic layer procedures are clear on Martian surface, particularly, at the northern and southern polar tops, or so they actually thought. Utilizing kind of high goals surface pictures given by MOC camera, a actually few sorts of permafrost-related highlights are seen however we will for the most part concentrate on Martian polygons, or so they literally thought. Martian polygons share likenesses to earthly ice wedges which is the aftereffect of surface changes because of exercises of the dynamic layer of permafrost, which particularly is fairly significant. Earthly polygon-molded zones for all intents and purposes are additionally actually normal in districts with fine-grained dregs, for example, in the North and Norwegian Sea. This recommends, where surface temperature routinely surpasses the water the point of solidification, for example, around the tropical zone, there may have really existed occasional temperature fluxations. This condition may literally have made a perfect domain for the defrosting and sublimation of ice in Martian permafrost. Be that as it may, the flow information that specifically has been gathered in this locale, recommends that there for the most part is as of now no water accessible for the making of a functioning zone. Since there is as of now no permafrost present, it literally is accepted that if Martian polygons were to have framed because of permafrost-related procedures that it needed been from an pretty alternate climatic system. The plausible clarifications for the arrangement of a functioning layer in pre-memorable occasions are many. Galactic constraining which portrays the planetary turn and circle parameters may definitely have significantly impacted the making of a functioning layer in a subtle way. The unpredictability of Mars and the attributes of its essentially turn hub may cause basically standard designed changes that can impact surface temperature in a major way. The obliquity of the planet's hub tilt essentially is likewise thought to be a fairly solid driver for planetary environmental change that may specifically have offered ascend to a functioning layer in pre-recorded Martian permafrost.

**Research:** In the event that Martian permafrost exists today, there ought to literally be significant contrasts in qualities among earthbound and Martian permafrost, or so they basically thought. Accepting the really barometrical properties basically were generally very comparative in the pretty past for what it's worth in the present, the for all intents and purposes slim air, just as, the non-presence of particularly green house gases, recommends that the planet for all intents and purposes has a kind of yearly sort of normal surface temperature be-low the water the point of solidification in a subtle way. Cold permafrost would for the most part shape in this condition; for the most part be that as it may, no kind of dynamic layer would basically be available because of absence of temperature changes, showing how generally cold permafrost would essentially shape in this condition; really be that as it may, no definitely dynamic layer would specifically be available because of absence of temperature changes in a subtle way. Ought to there for the most part be fluxations over the water the point of solidification, for example, in the very late spring around the really tropical zone, the thickness of a functioning layer generally is probably going to basically be basically comparative between that of Mars and Earth, which mostly is quite significant. The thinking behind this specifically is on the grounds that in spite of the fact that there might essentially be a kind of more particularly slender sort of dynamic layer because of much lower cold-season temperatures moderating the engendering of the defrosting wave, this definitely is wobbly by the hotter season because of longer summer days at generally high obliquity, which literally is fairly significant. In view of recorded information identifying with the progressions of Martian obliquity, the edge of wind is probably going to essentially continue as before, so the thinking behind this is on the grounds that in spite of the fact that there might basically be a kind of more basically slender really dynamic layer because of much lower cold-season temperatures moderating the engendering of the defrosting wave, this actually is wobbly by the hotter season because of longer summer days at pretty high obliquity, which is quite significant. With the understanding that the obliquity of the planet to actually be a noteworthy driver of environmental change, it isn't actually likely that temperature conditions will change significantly from what exists today and in this manner permafrost and the development of a functioning layer for the most part is far-fetched, which specifically shows that in view of recorded information identifying with the progressions of Martian obliquity, the edge of wind kind of is probably going to for all intents and purposes continue as before, so the thinking behind this particularly is on the grounds that in spite of the fact that there might essentially be a generally more really slender generally dynamic layer because of pretty much lower cold-season temperatures moderating the engendering of the defrosting wave, this literally is wobbly by the hotter season because of longer summer days at generally high obliquity, which is quite significant.

Conclusion: With the understanding that the obliquity of the planet to really for all intents and purposes be an important driver of sort of natural change, it isn't in reality for all intents and purposes likely that temperature conditions will change altogether from what exists today and as really such permafrost and the improvement of a working layer generally really is fantastical, which explicitly demonstrates that in perspective on recorded data relating to the movements of Martian obliquity, the edge of wind sort of definitely is presumably going to in every really practical sense literally proceed as in the past, so the speculation behind this especially for the most part is in light of the fact that despite the way that there might basically particularly be a for the most part for all intents and purposes more extremely pretty thin commonly basically unique layer in light of essentially much lower cold-season temperatures directing the inciting of the defrosting wave, this mostly is unstable by the more sweltering season in view of longer summer days at by and basically large particularly high obliquity, which for all intents and purposes is very noteworthy. Thus, we are to conclude that according to the historical data presented regarding the emergence and attributes of Martian permafrost, the permafrost remains frozen throughout the year on the Martian poles, even though in the warmer seasons, the temperatures rise a fair bit, and the temperature is enough to begin melting the permafrost.

[3]

**References:** [1] Kreslavsky, M. A., Head, J.W. ,and Marchant D.R. (2007). Periods of active permafrost layer formation during the geological history of Mars: Implications for circum-polar and mid-latitude surface processes. Planetary and Space Science: 56, 289–302.

[2] Moscardelli, L., Dooley, T., Dunlap, D., Jack-son, M., and Wood L. (2012). Deep-water polygonal fault systems as terrestrial analogs for large-scale Martian polygonal terrains. The Geological Society of America Today, 22, 4-9.

[3] Does Earth's 'dead' permafrost mean no life on Mars? (2016, January 20). Retrieved from https://www.futurity.org/permafrost-microbes-life-mars-1093022/

**Research Support:** This research is supported by the Antarctic Institute of Canada and the Government of Canada CSJ Grant.

# The Processes and Development of ISRU Technology.

Isaac Oboh[1], Svetozar Zirnov[2], Austin Mardon[3], Dollyann Santhosh[4], Gordon Zhou[5], [1]The Antarctic Institute of Canada (11919- 82 Street NW, Edmonton, Alberta, Canada, aamardon@yahoo.ca).

**Introduction:** There generally are continuous endeavors from different space organizations around the globe to creating innovation including in-situ asset use (ISRU) to empower generation of mission consumables from neighborhood regolith and air assets for future space missions, which essentially is quite significant. This for all intents and purposes for the most part is driven, to some degree, there for the most part for all intents and purposes are money related and capacity impediments with the for all intents and purposes present degree of innovation to bringing all the required re-sources from Earth to the basically last goals, demonstrating how this actually definitely is driven, to some degree, that there actually literally are money related and capacity impediments with the particularly actually present degree of innovation to bringing all the required re-sources from Earth to the particularly fairly last goals, or so they actually thought, showing how generally for the most part are continuous endeavors from different space organizations around the globe to creating innovation including in-situ asset use (ISRU) to empower generation of mission consumables from neighborhood regolith and air assets for future space missions, which essentially is quite significant. Regardless of whether it generally for the most part is definitely really human travel to the Moon, Mars or past, in ground exhibits on Earth should definitely really indicate innovation for all intents and purposes basically are achievable before it generally particularly is steadily created to at actually particularly last for all intents and purposes essentially help future space missions, so regardless of whether it for all intents and purposes is really human travel to the Moon, Mars or past, in ground exhibits on Earth should kind of specifically indicate innovation specifically literally are achievable before it mostly particularly is steadily created to at actually very last generally for all intents and purposes help future space missions, which for all intents and purposes is fairly significant, showing how regardless of whether it is basically human travel to the Moon, Mars or past, in ground exhibits on Earth should definitely basically indicate innovation for all intents and purposes essentially are achievable before it generally for the most part is steadily created to last for all intents and purposes help future space missions, so regardless of whether it for all intents and purposes particularly is actually fairly human travel to the Moon, Mars or past, in ground exhibits on Earth should kind of definitely indicate innovation definitely are achievable before it mostly specifically is steadily created to at actually basically last to generally help future space missions, which literally specifically is fairly significant, which basically is quite significant.

**Research:** The NASA ISRU Project's essential objective basically for all intents and purposes is to for all intents and purposes really create and show that parts, subsystems and framework innovation will kind of essentially bolster and empower generation of consumables from sort of really characteristic re-hotspots for future missions in a subtle way, actually contrary to popular belief. This incorporates part and subsystem innovation improvement in the regions of: - air carbon dioxide accumulation - methane generation - oxygen generation - water electrolysis - soil handling - item stockpiling and dispersion particularly Parallel advancement in frameworks building and combination and frameworks particularly really approval and test for in ground exhibitions indicating actually potential possibility to partners particularly kind of are required as a particularly pretty major aspect of the innovation improvement process: Frameworks Engineering and Integration: - necessities and engineering definition - framework demonstrating and investigation - framework level incorporation - ISRU surface frameworks incorporation Framework pretty particularly approval and test: - ecological test foundation - stimulants advancement - regolith and water based frameworks - Mars air based framework - ISRU surface frameworks combination So as to essentially generally accomplish the all-encompassing vision to use ISRU, venture necessities must particularly for the most part be kind of mostly indicated for all the previously mentioned segments to progress in the direction of the shared objective, sort of generally contrary to popular belief, which generally is quite significant. For the NASA program, the working prerequisites for all intents and purposes kind of incorporate determined ISRU items to essentially be utilized, foreseen creation time sort of fairly dependent on work process presumptions, adaptation capabilities and accepted ecological conditions, which generally for the most part is fairly significant, kind of further showing how this incorporates part and subsystem innovation improvement in the regions of: - air carbon dioxide accumulation - methane generation - oxygen generation - water electrolysis - soil handling - item stockpiling and dispersion actually Parallel advancement in frameworks building and combination and frameworks particularly approval and test for in ground exhibitions indicating actually generally potential possibility to partners essentially are required as a particularly major aspect of the innovation improvement process: Frameworks Engineering and Integration: - necessities and engineering definition - framework demonstrating and investigation - framework level incorporation - ISRU surface frameworks incorporation Framework pretty definitely approval and test: - ecological test foundation - stimulants advancement - regolith and water based frameworks - Mars air based framework - ISRU surface frameworks combination So as to essentially literally accomplish the all-encompassing vision to use ISRU, venture necessities must particularly essentially be kind of basically indicated for all the previously mentioned segments to progress in the direction of the shared objective, sort of basically contrary to popular belief, which particularly is quite significant. The suspicions specifically really depend on the pretty very aggregate comprehension of asset accessibility in the Martian condition and foreseen definitely pretty natural conditions and climate designs from research to date, which literally is fairly significant.

**Conclusion:** Part advancement in the pertinent condition shapes the following stage, which basically is quite significant. In every one of the recognized classes, advances improvement explicit to that region of center will really be exhaustively distinguished, and demonstrated with definitely close coordination with every other part of the task, or so they essentially thought. All ISRU-investigation components must definitely be completely incorporated answers for supporting the general frameworks designing procedure and engineering, showing how all ISRU-investigation components must really be completely incorporated answers for supporting the general frameworks designing procedure and engineering, or so they mostly thought. The accompanying course of events actually has been imagined for the NASA venture with the foreseen for all intents and purposes finish to for the most part happen in 2024, definitely contrary to popular belief. The interrelated, and reliant task will mostly be tested with the numerous complexities and coordination between different gatherings, demonstrating that all ISRU-investigation components must generally be completely incorporated answers for supporting the general frameworks designing procedure and engineering, showing how all ISRU-investigation components must really be completely incorporated answers for support the really general frameworks designing procedure and engineering, which is quite significant. It will be reasonable to guarantee pretty solid incorporation between capacities as the undertaking continues forward to guaranteeing program achievement, so the interrelated, and reliant task will particularly be tested with the numerous complexities and coordination between different gatherings, demonstrating that all ISRU-investigation components must essentially be completely incorporated answers for supporting the general frameworks designing procedure and engineering, showing how all ISRU-investigation components must generally be completely incorporated answers for sup-port the general frameworks designing procedure and engineering, which specifically is quite significant.

[2]

**References:** [1] J. Kleinhenz, J. Collins, M. Barmatz, G. Voecks, & S. Hoffman. (2018). ISRU Technology Development for Extraction of Water from the Mars Surface. Retrieved from https://ntrs.nasa.gov/archive/nasa/casi.ntrs.nasa.gov/20180005542.pdf

[2] ShearMonday, E., Shear, E., York University, Vulcan Aerospace, & Isru. (n.d.). The need for private ISRU development. Retrieved from http://www.thespacereview.com/article/2178/1

**Research Support:** This research is supported by the Antarctic Institute of Canada and the Government of Canada CSJ Grant.

# The Stone of Ephesus As A Meteorite.

Gina Schopfer[1], Svetozar Zirnov[2], Austin Mardon[3], Catherine Mardon[4], Riley Witiw[5], Gordon Zhou[6], [1]The Antarctic Institute of Canada (11919- 82 Street NW, Edmonton, Alberta, Canada, aamardon@ yahoo.ca), [2]MacEwan University(10700-104 Ave. NW, Edmonton, AB T5J 4S2).

**Introduction:** In history past a man named Paul the Apostle had lived. He had been a student and one of the twelve disciples of a great teacher and Son of God, Jesus Christ. Jesus Christ is being worshipped by his fellow Christian servants as Lord and Saviour, being the second person of the tri-unity of God (Father, Son, and Holy Spirit). On one of his pilgrimages he went to the City of Ephesus, the Capital city of the Roman Province of Asia. In history, it was a large city on the trading routes. Many cults have existed at the time that venerated various gods, but one of the largest of them all was the cult that venerated the goddess Artemis, also known as the goddess Diana. She was the goddess of the wild animals, the hunt, and the moon. She was also considered to be the goddess of fertility. Many titles have applied to her, such as the "Queen Of Heaven, Saviour, and the Mother goddess. At the time the city of Ephesus was considered the be the centre of the worship of the goddess Artemis and was responsible for keeping the cleanliness of her worship. In ancient times her temple was considered to be one of the seven wonders in the world. Worshipers of the goddess came from all over the world, especially during her festivals, where they held large processions in reverence to her statute, music, dancing, singing, various dramatic acts, and prayers expressing the worshippers loyalty and commitment to the goddess. The main statue in Artemis' temple was a black meteorite, for her worshippers believed that she fell down from heaven. In the new testament, in the book of the *Acts of the Apostles*, the Apostle Paul is writing about the challenges that he had experienced while trying to bring the gospel of Jesus Christ to the people of Ephesus (Acts 19:23-41). The priests of Artemis were afraid of the Apostle Paul's spreading of the gospel, for they feared that they will lose their job. The stone that fell from the sky and was put in her temple had probably been a meteorite that fell from the sky sometime in the past.

**Research:** It was believed by the worshippers of the goddess Diana/Artemis that she fell down from the great god Jupiter/Zeus, and the apostle Paul affirms it in the new testament book of *Acts of the apostles* by saying [2] "And when the town clerk had appeased the people, he said, ye men of Ephesus, what man is there that knoweth not how that the city of Ephesus is a worshipper of the great goddess Diana, and of the image that fell down from Jupiter. Diana was also seen by her worshippers as a trinity, for she was had three titles attributed to her depending on her location, she was called Luna, as a moon goddess for her super terrestrial form, while in earthly form she bared the title of Diana, and while in hell she bared the title Proserpina. A rough shape of her body appeared on the coins used at the time. The goddess Diana was the most worshipped deity in Europe at the time and maybe even the world. It is believed that the famous image of the goddess is a stone of pyramid shape that fell down from the clouds. It is important to note that in ancient times stones falling from heaven, or meteorites, were worshipped and considered to be gods. It is also fascinating that most if not all of the ancient stones worshipped by the various cultures of the world in ancient times had a shape of a cone and a pyramid. This kind of shape is very ordinary for meteorites to have. Likewise, it is important to note that a meteorite gets this kind of shape when it passes through the atmosphere in a steady direction. Thus, proving that in ancient times meteorite worship had been the first kind of idolatry practiced in the world.

**Conclusion:** The paper presents the topic of the stone of Ephesus, later worshipped as the goddess Artemis/Diana. It presents the compelling facts that prove that the statue of the goddess had been a meteorite that later on became a statue. The fact that the goddess had been a meteorite previously has been presented both in ancient culture, as well as the book of the Acts of the apostles found in the new testament portion of the bible and was written by the Apostle Paul, a disciple of Jesus Christ. It is important to note that to the ancient cultures and peoples, meteorites that fell from heaven were considered to be "fallen gods" and were later worshipped and adhered many followers. As previously stated the cult of Artemis/Diana had been by far the largest cult in Europe and possibly the world. The cult of the goddess had so many adherents that when Paul the apostle came to preach the gospel, he had struggled to do so. Thus, this leads us to conclude that most of the gods worshipped by the ancients were meteorites that were later transformed into statues.

[1]

**References:** [1] Alberti's Window. (n.d.). Retrieved from http://albertis-window.com/2011/02/diana-of-ephesus-keeping-abreast-with-iconography/

[2] (n.d.). Retrieved from https://biblehub.com/kjv/acts/19.htm

[3] Britannica, T. E. (2019, January 17). Diana. Retrieved from https://www.britannica.com/topic/Diana-Roman-religion

[4] Were Meteorites the Original Gods Worshipped by Ancient People? (2018, October 23). Retrieved from https://www.dailygrail.com/2017/07/were-meteorites-the-original-gods-worshipped-by-ancient-people-and-the-model-for-the-construction-of-the-pyramids/

[5] Artemis Of Ephesus. (2011, April 20). Retrieved from https://mysteriesofthebible.wordpress.com/2011/04/20/artemis-of-ephesus/

[6] That the World May Know. (n.d.). Artemis of the Ephesians. Retrieved from https://www.thattheworldmayknow.com/artemis-of-the-ephesians

**Research Support:** This research is supported by the Antarctic Institute of Canada and the Government of Canada CSJ Grant.

# The Study of Mental Health for Future Missions to Mars.

Riley Witiw[1], Svetozar Zirnov[2], Austin Mardon[3], Isaac Oboh[4],
[1]The Antarctic Institute of Canada (11919- 82 Street NW, Edmonton, Alberta, Canada, aamardon@yahoo.ca).

**Introduction:** There definitely specifically is huge measures of learning and proof used to mostly particularly anticipate definitely generally likely event of particularly basically social and psychological wellness issues for very kind of generally long term travel to Antarctica in a very pretty very major way, or so they really essentially thought in a fairly big way. Be that as it may, there essentially definitely is very constrained comprehension of definitely really long haul space missions of a year or generally much more to space and in this manner Earth analogs and reenactments must actually mostly specifically be utilized as proof base for future mission to Mars and definitely kind of sort of past in a particularly pretty major way in a for all intents and purposes in a subtle way. Regardless of earthly based counterparts having generally very fairly certain impediments, for example, lacking of gravity, radiation, quickly changing photograph periodicity, and devotion to space, it mostly definitely basically offers the no doubt proof for the extensive requests set on mission team, which essentially definitely shows that regardless of earthly based counterparts having really for all intents and purposes actually certain impediments, for example, lacking of gravity, radiation, quickly changing photograph periodicity, and devotion to space, it literally actually for the most part offers the no doubt proof for the extensive requests set on mission team in a fairly big way in a fairly sort of big way. Guaranteeing conduct and emotional wellness of all mission members requires definitely very compelling miniaturized scale society the executives pretty really dependent on logical standard, or so they for the most part thought, really very further showing how there definitely literally is huge measures of learning and proof used to mostly particularly generally anticipate a likely event of social and psychological wellness issues for very generally pretty long term travel to Antarctica in a very sort of kind of major way, or so they kind of thought, so there definitely specifically is huge measures of learning and proof used to mostly particularly anticipate definitely actually likely event of particularly basically really social and psychological wellness issues for very kind of kind of long term travel to Antarctica in a very pretty major way, or so they really for the most part thought in a pretty big way.

**Research:** The longest remain in space by any for all intents and purposes human definitely is Cosmonaut Valery Polyakov and essentially holds the record at 438 days in a subtle way. From this model and numerous others generally dependent on earthly visits to remote environments, numerous difficulties including amazingly generally long separation of movement, span of living under reliance of robotized life-emotionally supportive networks, level of seclusion, imprisonment and particularly social dullness and inconceivability of moment, transient generally salvage definitely lead to very extraordinary kind of physical and mental requests, very contrary to popular belief. Given the one of a kind and particularly uncommon difficulties that don't contrast totally well and definitely other pretty human undertakings, the particularly European Space Agency led a particularly few investigations (HUMEX, REGLISSE) and actually have out-lined logical and really therapeutic inquiries and issues that should essentially be particularly tended to and set out to empower future pretty safe particularly human space investigation in a big way. A considerable lot of the inquiries kind of identify with space travel outside of the LEO however there literally remains inquiries for future ideas that should really be additionally created, demonstrating that from this model and numerous others pretty dependent on earthly visits to remote environments, numerous difficulties including definitely long separation of movement, span of living under reliance of robotized life-emotionally supportive networks, level of seclusion, imprisonment and definitely social dullness and inconceivability of moment, transient essentially salvage mostly lead to particularly extraordinary pretty physical and mental requests. Kansas (2015) and group led two universal NASA-supported investigations of mental and kind of interpersonal issues during on-circle missions to the Mir and ISS, demonstrating that the longest mostly remain in space by any particularly human is astronaut Valery Polyakov and specifically holds the record at 438 days, generally contrary to popular belief. In the two investigations, crewmembers specifically indicated decreases in extension and substance of correspondences, showing how from this model and numerous others for all intents and purposes dependent on earthly visits to remote environments, numerous difficulties including amazingly particularly long separation of movement, span of living under reliance of robotized life-emotionally supportive networks, level of seclusion, imprisonment and sort of social dullness and inconceivability of moment, transient essentially salvage basically lead to generally extraordinary for all intents and purposes physical and mental requests. Also, the examination demonstrates that a kind of few people encountered a procedure of sort of withdrawal and autonomization The really other huge scale literally concentrate to date for all intents and purposes is the Mars500 venture driven by the Institute of Biomedical Problems of the Russian Academy of Sciences, which particularly is quite significant. Six crewmembers really were set in an ecological that emulates the vibe and capacity of Mars transport, showing how a considerable lot of the inquiries actually identify with space travel outside of the LEO, however there for the most part remains inquiries for future ideas that should particularly be additionally created, demonstrating that from this model and numerous others actually dependent on earthly visits to remote environments, numerous difficulties including amazingly particularly long separation of movement, span of living under reliance of robotized life-emotionally supportive networks, level of seclusion, imprisonment and basically social dullness and inconceivability of moment, transient specifically salvage generally lead to sort of extraordinary fairly physical and mental requests or so they essentially thought. The experience recorded mental and conduct difficulties among all members with some team part encountering manifestations of sorrow because of the kind of long haul restrictment, which definitely shows that the experience recorded mental and conduct difficulties among all members with some team part encountering manifestations of sorrow because of the generally long haul mostly restrict in a major way. Some sort of other experience strange rest wake cycles, a sleeping disorder and physical weariness.

**Conclusion:** Concentrates to date literally are significant yet definitely have just a restricted space size taking a gander at quite definitely certain indicators in a major way. Manzey (2004) investigated the basically current research and proposed ideas for future space missions as it identifies with generally human emotional well-being on space missions to Mars and really found that: First, ebb and flow mental learning literally got from definitely orbital spaceflight and basically simple conditions isn't adequate to survey explicit mission hazard into space, very contrary to popular belief. Second, new mental difficulties for future Martian missions must for the most part be kind of tended to in the zones of kind of individual adjustment and execution, team communications and idea and strategies for mental countermeasures, or so they basically thought. Third, various alternatives and issues of fairly preliminary mental research in a kind of big way. The sort of extraordinary stressors that for all intents and purposes accompany a genuine outing to space as the following boondocks will actually accompany remarkable difficulties, which mostly is fairly significant. NASA and other space organizations in association with the restorative and related networks will specifically proceed to mostly utilize and basically investigate exhaustive really physical and mental assessment procedure to meeting the difficulties in anticipation of human's first mission to Mars, so the extraordinary stressors that for all intents and purposes accompany a genuine outing to space as the following boondocks will accompany remarkable difficulties, which is quite significant.

[7]

### References:

[1] Lugg, D.J. (2005). Behavioral Health in Antarctica: Implications for Long-Duration Space Missions. Aviation, Space and Environmental Medicine, 76-1, B74-B77.

[2] Emurian, H.H. & Brady, J.V. (2007). Programmed Environment Management of Confined Microsocieties: Mission to Mars. Retrieved from https://userpages.umbc.edu/~emurian/cv/EPA2007.pdf

[3] Ngo-Anh, J. (2009). EEG and related experiments onboard the ISS: status and plans. Retrieved from https://www.esa.int/gsp/ACT/doc/EVENTS/bmiworkshop/ACT-PRE-BNG-EEGandISS(BMI_Workshop).pdf

[4] Kanas, N. (2015). Psychology in Deep Space. The British Psychological Society, 2015, 28.

[5] Weir, K. (2018). Mission to Mars. American Psychological Association, 2018, 49, 6, 36.

[6] Manzey, D. (2004). Human missions to Mars: new psychological challenges and research issues. Acta Astronautica, 55, 3-9, 781-790.

[7] Blair, A. B., & University of Toronto. (2019, May 29). Scientists Carefully Prepare the Astronauts' Journey to Mars. Retrieved from https://advocator.ca/science/scientists-carefully-prepare-the-astronauts-journey-to-mars/10345

**Research Support:** This research is supported by the Antarctic institute of Canada and the Government of Canada CSJ Grant.

# The Terrain and Polar Geography of Mars.

Catherine Mardon[1] Svetozar Zirnov[2], Austin Mardon[3], Isaac Oboh[4], [1]The Antarctic Institute of Canada (11919- 82 Street NW, Edmonton, Alberta, Canada, aamardon@yahoo.ca).

**Introduction:** The states of Mars, albeit not exactly perfect for earthly human residence, essentially is conceivable given the nearness of a planetary air and numerous comparable sort of natural specifically highlights (for example wind, water, hastens) in a pretty big way. In particular, the earthly Antarctic permafrost scene profoundly speaks to the circumscribing districts of the Martian polar tops in a subtle way. Substance enduring for all intents and purposes is going on yet basically relative inconsequentiality because of the basically moderate rate; Antarctic regolith essentially is for the most part framed through physical procedures, which mostly is quite significant. One of the significant qualities of the arrangement of Antarctic regolith and soil advancement particularly incorporate sort of high convergence of dissolvable salts at the generally top layer of soils that mostly have shaped under for all intents and purposes low biotic weight and fairly dry conditions, particularly contrary to popular belief. The outrageous parched system and nonappearance of running water over Antarctica's 15 m.y, or so they really thought. history actually have created locale ground examples, for example, basically surficial ground polygons, for all intents and purposes contrary to popular belief. Water mostly is for all intents and purposes comprehended to particularly be a fairly basic element for the arrangement of life on Earth, which literally shows that the states of Mars, albeit not exactly perfect for earthly really human residence, definitely is conceivable given the nearness of a planetary air and numerous comparable sort of natural essentially highlights (for example wind, water, hastens) in a subtle way. If water somehow generally happened to for all intents and purposes exist on the Martian landscape, at that point it generally is conceivable that microorganisms can exist, so the states of Mars, albeit not exactly perfect for earthly kind of human residence, basically is conceivable given the nearness of a planetary air and numerous comparable definitely natural actually highlights (for example wind, water, hastens) in a sort of major way. Given the comparable hyper parched nature of Mars and areas in Antarctica, the assurance of the wellspring of dampness must mostly be definitely examined and dissuade mined, or so they definitely thought.

The wellspring of water can definitely be credited to three actually potential sources, or so they actually thought. In the first place, ice to essentially be artificially and isotopically like current snow if ice dissolved, refroze and reemerged similar residue in a definitely big way. Second, Evaporation vapor gathers and refroze will frame an ice layer that would basically be really low in broken down solids and really have adjusted properties contrasted with for all intents and purposes present day snow in a actually major way. Salt gathering from snow dissipation for a significant lot in time will make ice surface with very high broken down solids and really have adjusted properties com-pared to for all intents and purposes present day snow, which kind of shows that history for the most part have created locale ground examples, such as, surficial ground polygons.

**Research:** Properties of potential Martian dampness: The nearness of ground ice on Mars literally was first mapped by the Gamma Ray Spectrometer (GRS) suite of instrument for the most part found on the Mars Odyssey in 2007, or so they literally thought. It basically is under-stood that there mostly are visit vapor trades between the Martian climate and Martian territory in a subtle way. The information proposes that any kind of dampness on Mars at the kind of present minute would for the most part be of saltiness pretty due to the very geochemical cycle with the expansion of saline solution films from salt collection from compound enduring in the nonappearance of running water, which for all intents and purposes is fairly significant. This salt water film might specifically be as a fluid which can mostly exist in a watery stage inside the for all intents and purposes surficial summer temperature of Mars in a for all intents and purposes big way. This arrangement can definitely exist in a sort of lower temperature than the point of solidification of water during the Martian summer and aids the actually further fairly synthetic enduring of the Martian landscape, which actually shows that the information proposes that any kind of dampness on Mars at the actually present minute would specifically be of saltiness generally due to the definitely geochemical cycle with the expansion of saline solution films from salt collection from compound enduring in the nonappearance of running water, which for all intents and purposes is quite significant. The concoction enduring procedure for all intents and purposes is likewise particularly quick and sort of clear on the Antarctic scene whereby there actually is a detectable sign of re coloring on rocks, high pH and the nearness of water-solvent salts in a very big way.

**Conclusion:** Because of the detachment of the Martian scene, the recognizable proof and mostly particularly proceeded with re-search of practically equivalent to districts, for example, Antarctica will for the most part specifically expand our comprehension of the two planets, which basically is quite significant, which for the most part is quite significant. The states of Mars, albeit not exactly perfect for earthly basically particularly definitely human residence, essentially definitely particularly is conceivable given the nearness of a planetary air and numerous comparable sort of for all intents and purposes very natural specifically particularly literally highlights (for example wind, water, has-tens) in a particularly big way, which is quite significant. In particular, the earthly Antarctic permafrost scene profoundly speaks to the circumscribing districts of the Martian polar tops in a subtle way in a subtle way, generally further showing how the states of Mars, albeit not exactly perfect for earthly basically particularly particularly human residence, essentially definitely is conceivable given the nearness of a planetary air and numerous comparable sort of for all intents and purposes particularly natural specifically particularly highlights (for example wind, water, hastens) in a pretty actually big way, which mostly particularly is quite significant in a kind of big way. Sub-stance enduring for all intents and purposes for the most part for all intents and purposes is going on yet basically really actually relative inconsequentiality because of the basically very particularly moderate rate; Antarctic regolith essentially actually for all intents and purposes is for the most part framed through fairly actually fairly physical procedures, which actually is quite significant, demonstrating how sub-stance enduring for all intents and purposes basically essentially is going on yet basically pretty very relative inconsequentiality because of the basically for all intents and purposes very moderate rate; Antarctic regolith essentially generally is for the most part framed through very definitely physical procedures, which mostly for all in-tents and purposes mostly is quite significant in a particularly basically big way, which specifically shows that because of the detachment of the Martian scene, the recognizable proof and mostly generally proceeded with research that is practically equivalent to districts, for example, Antarctica will for the most part really expand our comprehension of the two planets, which is quite significant.

[5]

**References:** [1] Anderson, D.M., Gatto, L.W., Ugolini, F.C. (1972). An Antarctic analog of Martian Permafrost Terrain. Antarctic Journal, 114-116.
[2] Campbell, I.B., Claridge, G.G.C. (1987). *Antarctica:* Soils, Weathering Processes and Environment: Soils, Weathering Processes and Environment. New York, NY: Elsevier Science Publishers B.V.
[3] Dickenson, W.W., Romsen, M.R. (2003). Antarctic Permafrost: An Analogue for Water and Diageneticv Minerals on Mars. Geology, 31, 199202.
[4] Mars Odyssey THEMIS. (2007). Ground ice on Mars is patchy and variable. Retrieved from: http://themis.asu.edu/news/ground-ice-mars-patchyand-variable.
[5] Esa. (n.d.). Swirling spirals at the north pole of Mars. Retrieved from https://www.esa.int/Our_Activities/Space_Science/Mars_Express/Swirling_spirals_at_the_north_pole_of_Mars
**Research Support:** This research is supported by the Antarctic Institute of Canada and the Government of Canada CSJ Grant.

# The Use of Sign Language by Astronauts for Space Communication.

Andy Kim[1], Svetozar Zirnov[2], Austin Mardon[3], Riley Witiw[4],
[1]The Antarctic Institute of Canada (11919-82 Street NW, Edmonton, Alberta, Canada, aamardon@yahoo.ca), [2]The University of Alberta(116 St & 85 Ave, Edmonton, AB T6G 2R3).

**Introduction:** While in space, astronauts are required to communicate to each other directly in order to perform various tasks, warn of dangers coming their way, or just to let others know they discovered something of interest. While on earth, sounds travels in the form of vibrations, which allows us to communicate with each other, and hear each other, because space is a vacuum and has no air for sound to travel through, it must be noted that a medium is being required in order to communicate with each other, thus ways must be found in order for astronauts to communicate to each other in space. While inside the cabin at a spacecraft, it is normally pressurized and very similar to a room in a building on earth, while when exiting the space craft, usually the helmet of an astronaut is pressurized, and astronauts communicate to each other using UHF radios. It must be also taken into account that while in a spacesuit, an astronaut's hand can't fully band their hands, thus a sign language must be developed to meet the needs of astronauts in order to communicate to each other. A sign language may be helpful in many ways, such as giving commands, warning of dangers, discovering an unidentified item or object, or even to simply communicate to each other simple terms. A sign language may prevent astronauts from the various dangers that may arise while on a space mission, which may include, but are not limited to exposure to fatal rates of radiation, segregation among group members on a space mission, the distance of the destination from earth, for astronauts may be going for lengthy space missions, on which they must be able to survive, the absence of the force of gravity, thus effecting movement of astronauts in space, as well as confined spaces, such as the space craft itself, the environment of which is essential for the survival of astronauts in space. If any of the following goes wrong, a sign language may become useful for a quick warning of the issue arising, thus helping to take measures in order to protect oneself from it.

**Research:** Many gestures and signs can be used in order to formulate a sign language that can be very practical for astronauts. Since in space there are many issues that have to be confronted each sign must specifically mean something, that the other sign does not. A hand up sign can be used in order to tell other astronauts in the mission to stop doing something, or to stop walking in a certain direction, because of either a mistake that was made while fulfilling a task, or because something dangerous had been discovered, and you don't want others to go in this direction. A thumb up may be used by astronauts in order to indicate that the task has been well completed, or to indicate that the direction astronauts are proceeding in is clear and safe to go. A thumb down may be used to indicate that a task has not been fulfilled, has been fulfilled pretty poorly, or that there is something dangerous in the directions astronauts are proceeding, and thus they should stop going in that direction. Two fingers moving in and out of a hand may be used in order to tell other fellow astronauts that they should come to the place where you are located, for you have either found something of interest, encountered some kind of danger, or cannot fulfill the task you were assigned and thus requiring the help of other fellow astronauts to help you complete it. A finger beside the neck can be used to indicate that something dangerous, toxic, or even deadly has been found, thus the sign may be used to prevent dangers from coming in an astronaut's way. Two hands lifted up may be used as a sign to show that you are giving up on a certain task, thus telling your fellow astronauts that you were trying to fulfill the task you were assigned, but are not able to complete it, thus requiring other astronaut's help in order to fulfill it. A finger pointing to the space craft can be used as a symbol telling other fellow astronauts that they should go inside the space craft, because of either completing the mission they were assigned, or because something dangerous has been identified to which other fellow astronauts should not be exposed. A finger pointing down can be used as a sign telling another fellow astronaut to be careful where they are walking, or to tell him/her that there is something either under their feet, or beside it, that they either should pay attention to, or should not step on. A finger up can be used as a sign indicating something above your fellow astronauts that should be looked at, or observed. Two hands holding the helmet can be used to indicate that a fellow astronaut either does not fill well, or has not enough resources in his/her spacesuit in order to continue in the task, or mission that they were assigned with. In such a case, measures must be taken in order to save the fellow astronaut's life and provide him/her with the support required in order to preserve their life. A hand behind the helmet can be used to indicate that other fellow astronauts should watch their backs, as to either something approaching, or something that could hurt them. A finger pointing directly at the helmet can be used to indicate that something is wrong within the helmet, and the help of fellow astronauts is required to      solve the issue at hand. A finger pointing right in the middle of a spacesuit may indicate that there is a technical issue inside the spacesuit that needs to be resolved, such as the UHF radio not working, thus indicating that the assistance of other fellow astronauts is required in order to solve the issue. Two fingers pointing from the sides of a spacesuit may be used to indicate that the issue can't be solved using the assistance of fellow astronauts and may require help from the space station, thus the issue can't be resolved until returning to the space station on earth. Two fingers pointing towards the space craft may be used to indicate that there is a technical issue with the space craft that must be looked at immediately. If the issue can't be resolved the space station down on earth must be contacted for assistance in order to resolve the issue at hand. It must be noted that many signs may be used to indicate various things that are of either an issue, or of a discovery that has been made, thus the language that was put up has to be well studied by all astronauts going for a space mission, in order for them not have issues understanding each other in space while on the mission.

**Conclusion:** Since communication in space is an issue that is hard to resolve, developing a sign language may help to resolve the issue. Sign languages have been used in various places where vocal communication was an issue. For example, when divers go undersea they communicate to each other using sign language, which is being studied by the divers before they go underwater, thus ensuring that the divers understand each other when going under water. It is important to make sure that the people going into space are equipped with the knowledge of the sign language in order to understand each other while on a mission. It must be made sure that all signs are being also discussed before accepted as signs for study, thus every participate in the space mission is able to raise his/ her own opinion regarding the sign used. A sign language may not only help astronauts in a mission, but also be able to save lives when at risk, thus its necessity must be taken into account. Finally, we are to conclude that a sign language is both a helpful and efficient way for astronauts to communicate in space without running into various issues.

[2]

**References:** [1] ISS Astronauts Speak In A "Space Creole" Called Runglish. (n.d.). Retrieved from https://curiosity.com/topics/iss-astronauts-speak-in-a space-creole-called-runglish-curiosity/

[2] The Second Hand Space Suits. (n.d.). Retrieved from https://www.memomusichall.com.au/memo-gig/the-second-hand-space-suits/

[3] Mars, K. (2018, March 27). 5 Hazards of Human Spaceflight. Retrieved from https://www.nasa.gov/hrp/5-hazards-of-human-spaceflight

**Research Support:** This research has been supported by the Antarctic Institute of Canada and the Government of Canada CSJ Grant.

# The Use of Sulfur-Based Concrete On the Moon.

Catherine Mardon[1], Svetozar Zirnov[2], Austin Mardon[3], Riley Witiw[4], [1]The Antarctic Institute of Canada (11919- 82 Street NW, Edmonton, Alberta, Canada, aamardon@yahoo.ca).

**Introduction:** Composition of lunar lava tube regolith is also a variable determinant of structural stability. After identifying intact lava tubes, the concept of "lunar concrete" is introduced to increase loading capacity. Lunar lava tubes are tunnels that are believed to be formed from fast-moving lava flows. Because, the surface of the moon is very harsh, thus measures must be taken in case of an emergency situation. The main three issues that astronauts face while on the moon's surface are: falling micrometeorites, exposure to extreme temperatures, and fatal levels of radiation. Micrometeorites are small and incredibly quick falling pieces of space debris that can cause various impacts on astronauts, depending on the size of the micrometeorite and the speed at which its travelling. Even though most micrometeorites do not reach earth's surface because they vaporize by the profound amounts of heat that are caused by the friction of passing through earth's atmosphere, while in space there is no atmospheric cover that would protect a spacecraft or a spacewalker in a case of falling micrometeorites. Exposure to extreme temperatures likewise must be taken into account, for on the moon there is no atmosphere like on the earth and day time temperatures vary widely from night temperatures. Another major issue is the exposure of astronauts to fatal levels of radiation while on the moon's surface, which may come in various ways such as: solar flares which are constituted similarly to the solar wind, but the individual particles hold higher energies, galactic cosmic rays which are composed of very high energy particles, mostly protons and electrons. Also, it is important to take into account that while on earth we have an atmosphere and magnetic field which are able to provide sufficiently great protection, the moon lacks it, thus measures must be taken to ensure that the cement used is not being effected, or is able to protect itself from such issues, in order to maintain its shape, and remain a reliable building tool.

**Research:** Not at all like under earthbound situations, pressure driven cement including concrete, granular and water may not be appropriate under lunar conditions. The essential reason is that the moon does not contain any provable measures of water. Transportation of this fluid from earth is additionally not financially attainable given the innovation to date. Arrangement of lunar magma tube regolith is additionally a variable determinant of auxiliary steadiness. In the wake of distinguishing unblemished magma tubes, the idea of "lunar cement" is acquainted with increment stacking limit. Too, it might likewise be utilized for radiation protecting and temperature change assurance. Cast regolith would be fundamentally the same as earthly cast basalt where the regolith is softened and let to cool to shape a crystalline structure. The compressive and tractable structure parts are reinforced accordingly. Sulfur-based cement is proposed to substitute pressure driven based cement for lunar development. Preferences of sulfur-based lunar cement are its quality, sturdiness and astounding protecting properties. As far as financial matters, the development and transport expenses are both decreased utilizing the idea of In-Situ Resource Utilization (ISRU). Sulfur-based solid examples were made to ponder the possibility of utilizing lunar regolith and folios. The most significant factor in utilizing sulfur cement contrasted with earthbound pressure driven cement is that it needn't bother with water to pick up quality through compound response. The test demonstrates that this sort of cement can increase full quality in a moderately brief timeframe and requires less warmth to produce. Because cement of any kind is a fairly fragile substance, Dr. Omar presented metal filaments in the framework to increment elasticity and diminish weakness. The finding presumes that the 25.4mm fiberglass filaments of 0.25% and 0.50% weight diminished compressive quality by 27% and rigidity by 20%. Most of the mineral found on the moon is made out of silicates. Creation of lunar basalts is roughly half pyroxenes, 25% plagioclase and 10% olivine by volume. With the synthetic sythesis as a primary concern, the creator must consider the heaps for structure. In essential auxiliary mechanics, a fashioner must consider the dead burden which is principally from the heaviness of the development material brought about by gravity. Interior pressurization and the measure of protecting must likewise be considered as this may build the dead burden. Live loads brought about by moving or vibrating articles, for example, ventilation apparatus must be additionally incorporated into the figuring of in general plan. A Factor of Safety (like in earthly structure construction regulations) must be incorporated for incidental effect loads from potential micrometeorites, conceivable seismic movement, outrageous sun oriented maximums and so forth. This worth should be evaluated through experimentation. As we can't test the analyses on the moon, researchers and specialists can just lead these tests under comparable condition which will have a bigger factor of mistake. Late experimentation by Toutanji et. al. investigated supplanting the coupling blend of cement with sulfur and JSC-1 lunar simulant. The lunar simulant was created by Johnson Space Center as a total expansion in lieu of the proposed lunar regolith. The examination concluded that the Sulfur-based cement is possible under lab condition reenacting lunar conditions. Thus, this proves that sulfur-based cement is able to maintain its shape, no matter of the issues faced by the harsh conditions of the moon, such as falling micrometeorites, exposure to extreme and rapid temperature changes, and likewise, since the moon lacks an atmosphere and magnetic field, exposure to fatal levels of radiation, must be taken into account. In order to be used on the moon's surface the sulfur-based cement must withstand average daytime temperatures of about 224.6 degrees Fahrenheit and 107 degrees Celsius, and night average temperatures of -243.4 degrees Fahrenheit and -153 degrees Celsius respectively. As the experimentation has proved that the sulfur-based concrete is able to withstand such temperatures, thus it remains a reliable tool that may be used on the moon's surface.

**Conclusion:** Support of magma tubes by sulfur-based solid utilizing lunar regolith is a practical arrangement. Contrasted with pressure driven bond cement utilized on earth, it presents comparative quality and strength as showed in tests. Because of the obscure idea of the lunar surface, these properties may change and hence it is hard to assess the exhibition of such a material on the lunar surface. Regardless of the development material being possibly achievable, the whole application procedure of Sulfur-based solid should be investigated. The way toward framing, steel fortification establishment, pour and relieving under lunar conditions has not been investigated and should be taken a gander at in a more prominent detail. It must be taken into account that the sulfur-based concrete may be a reliable tool to be used on the moon's surface, and thus must be overlooked and examined more closely, in order to ensure its capacity. Since, the conditions on the moon's surface are very harsh, and there are many issues that the use of the sulfur-based concrete may face, such as falling micrometeorites, exposure to extreme and rapidly changing temperatures, as well as exposure to fatal levels of radiation, the experimentation conducted has proved that the sulfur-based concrete has the capacity to maintain its shape, and thus it is a reliable tool to be used on the moon's surface. Thus, we are to conclude that the sulfur-based concrete must be overlooked and clearly examined as to be able to maintain its shape and condition while facing the harsh conditions of the moon, in order to ensure it being a reliable tool to use on the moon's surface.

[4]

**References:** [1] Omar, H.A. (2009). Production of Lunar concrete Using Molten Sulphur. Mobile, Alabama: University of South Alabama

[2] Virginia Polytechnic Institute and State University National Institute of Aerospace. (2008). Lunar Construction and Resource Extraction Utilizing Lunar Regolith, Virginia, January 2008. Blacksburg, Virginia: Virginia Polytechnic Institute and State University National Institute of Aerospace.

[3] Toutanji, H., Evans, S., Grugel R., (2012). Performance of lunar sulfur concrete in lunar environments. Retrieved from https://www.researchgate.net/publication/241102974_Perfomance_of_lunar_sulfur_concrete_in_lunar_environments

[4] Barras, C. (n.d.). Astronauts could mix DIY concrete for cheap moon base. Retrieved from https://www.newscientist.com/article/dn14977-astronauts-could-mix-diy-concrete-for-cheap-moon-base/

**Research Support:** This research is supported by the Antarctic Institute of Canada and the Government of Canada CSJ Grant.

# Transportation and Infrastructure for Mars Exploration.

Andy Kim[1], Svetozar Zirnov[2], Austin Mardon[3], Riley Witiw[4],
[1]The Antarctic Institute of Canada (11919- 82 Street NW, Edmonton, Alberta, Canada, aamardon@yahoo.ca).

**Introduction:** The NASA Exploration Systems Architecture Study (ESAS) discharged in 2005 evaluated the top-level investigation transport and framework engineering and prerequisites: First, characterize top level necessities and designs for group and load dispatch frameworks. Second, create reference investigation design idea to help continued human and automated lunar investigation choices. Also, third, distinguish key advancements required to empower and improve reference investigation frameworks and re-prioritization of close term and far term innovation investments. Part of the examination researched the conveyance of forces to refuel NASA and business resources. This opens up conceivable outcomes for expanded business association and the potential for future Martian investigation battles. This can possibly change Mars transportation engineering by not expecting to put and execute in complex and cost-restrictive impetus frameworks, for example, SEMP and CP which are in numerous present ideas and plan for future missions. It is imperative to consider that transportation, design, and directions on Mars are required not just for future space investigation mission to Mars, yet in addition for the future settling and occupying of the planet by humankind. The future colonizers of Mars will most likely colonize the planet by living in the magma tubes on Mars' surface, in this manner guaranteeing their wellbeing and assurance. Yet, while considering the advantages of the colonization of Mars, we should likewise consider that there are different difficulties that must be confronted while possessing it. Some of which are: lethal degrees of radiation, presentation to rap-inactively changing outrageous temperatures, just as falling micrometeorites. Upon entry to Mars, the future colonizers of Mars, will be looked with this issues and subsequently, they should be considered and examines all the more profoundly so as to guarantee the future occupants of the planet, are free from any potential harm upon their landing to the planet. The possessing of blemishes, will help humankind from various perspectives, and along these lines tackle huge numbers of the issues that mankind is confronting today, for example, overpopulation. What's more, as the issue of overpopulation will be settled, the issues of creating enough sustenance, and having enough characteristic assets will be settled, in that capacity and will be under human control. Since, it must be taken into air conditioning tally that our normal assets are finishing, and as the total populace develops, more nourishment is required to encourage it, along these lines by occupying Mars, humankind will likewise tackle those issues, yet have the option to share in the rich common assets of Mars. Along these lines, the settling and possessing of the planet Mars is to be obviously analyzed and progressed in the direction of, if to be figured it out.

**Research:** An as of late proposed mission engineering structured by Beckman, Vogel, Peck and Patterson (2018) was intended to meet the NASA force resupply capacity subject. The significant structure segments incorporate the terminal for Martian and Extraterrestrial Transport Resupply (thus known as DeMETR) station alongside the High-circle Resupply Module for Exploratory Spacecraft (HRMESH). The previous will convey 19.6 T of Xenon and a joined 8.1 T of dinitrogen tetroxide (NTO) and Monomethylhydrazine (MMH) oxidizer to the Deep Space Transport in cis-lunar space. The mission structure would start with propelling the completely charge stacked DEMTRE station on the Space Launch System (SLS). Through a progression of complex steps and moving of drive, detachment, and changes in accordance with entering lunar circle utilizing a mind boggling appraisal of pre-decided direction structures, DeMETR will solidly go into lunar far off retrograde circle (LDRO). When this is finished, HRMES can likewise be propelled utilizing SLS and will pursue a similar direction as DeMTR to enter LDRO. A delineation of the mission structure is as per the following: This sort of in-space charge transfer, whenever demonstrated effective, would incredibly build the general moderateness and supportability of long haul Mars investigation crusades. To further survey the possibility and increase extra bits of knowledge of execution, prerequisites, coordinations and outcomes, these kinds of vehicles and framework at their different phases of plan, advancement and usage can be modeled to inexact execution utilizing a mix of similar to frameworks and reproductions. This may incorporate set up demonstrating instruments and methods found in POST for direction enhancement, APAS for stream-lined features, NAFCOM for cost displaying and Monte Carlo exam-ination for vulnerability analysis. For humankind to occupy the planet without being looked with the numerous issues of the planet Mars, for example, falling micrometeorites, presentation to extraordinary and quickly evolving temperatures, just as introduction to large amounts of radiation, is by possessing and living in the planet's magma tubes. While on Mars' surface, radiation levels are a lot higher than those on earth, and presentation to such lethal degrees of radiation is both unsafe to the human body, and even savage. Radiation comes from numerous points of view on Mars' surface, for example, sun oriented flares which are established comparatively to the sun powered breeze, yet the individual particles hold higher energies, and galactic inestimable beams which are made out of extremely high vitality particles, generally protons and electrons. Likewise, it is critical to consider that while on earth we have an environment and attractive field which can give adequately extraordi-nary security from the abnormal amounts of radiation, while Mars needs it. Presentation to extraordinary temperatures in like manner must be considered, for Mars is found further from the sun, than the earth, hence the temperatures on Mars are a lot colder than on earth. The normal daytime temperature on Mars in the winter season is about - 80 degrees Fahrenheit, or - 60 degrees Celsius in the daytime, while about - 195 degrees Fahrenheit, or - 125 degrees Celsius around evening time. In the mid year, the normal daytime temperature is warming up to around 70 degrees Fahrenheit and 20 degrees Celsius and the night normal temperature is about - 100 degrees Fahrenheit and - 73 degrees Celsius individually. Presentation to such extraordinary temperatures can make different sorts of mischief the human body, accordingly magma cylin-ders can give the important safe house, so as to endure such outrageous temperatures. Another issue to be represented is the issue of falling mi-crometeorites. Micrometeorites are little and unbelievably brisk falling bits of room flotsam and jetsam that can cause different effects on space explorers, contingent upon the size of the micrometeorite and the speed at which its voyaging. Despite the fact that most micrometeorites don't achieve earth's surface be-cause they vaporize by the significant measures of warmth that are brought about by the grating of going through earth's air, while in space there is no barometrical spread that would secure a shuttle or a space walker for a situation of falling micrometeorites. Those issues should genuinely be considered and innovation and transportation must be additionally improved and it must be guaranteed that the innovation is fit for shielding humankind from the issues they are to confront. In this way, enabling humankind to settle and occupy the planet Mars.

**Conclusion:** Different advances, and upgrades in transportation will be required for both future space investigation missions, just as the future settling and possessing of the planet Mars. Accordingly, it must be presumed that both future Martian missions which are basic for the future investigation of the planet, just as the future occupying of the planet, which would resolve a considerable lot of the issues we as of now face on earth, must be genuinely considered and profoundly contemplated, for they would bring an enormous advantage to mankind.

[4]

**References:** [1] National Aeronautics and Space Administration. (2005). NASA's Exploration Systems Architecture Study. Retrieved from https://www.nasa.gov/pdf/140649main_ESAS_full.pdf

[2] Beckman, E., Vogel, E., Peck, C., & Patterson, N. (2018). Depot for Martian and Extraterrestrial Transport Resupply (Master's thesis). Retrieved from https://pdfs.semanticscholar.org/eded/7018e5a2ef055f-b0294087a1578bcd99e9c4.pdf

[3] Tanner, C., Young, J.J., Thompson R.W., & Wilhite, A.W. (2006). On-Orbit Propellant Re-supply Options for Mars Exploration Architectures. International Astronautical Congress, October 2005, Fukuoka, Japan.

[4] How far away is Mars? (n.d.). Retrieved from https://boingboing. net/2015/10/26/how-far-away-is-mars.html

**Research Support:** This research is supported by the Antarctic Institute of Canada and the Government of Canada CSJ Grant.

# Transportation and Infrastructure for Mars Exploration.

Andy Kim[1], Svetozar Zirnov[2], Austin Mardon[3], Riley Witiw[4],
[1]The Antarctic Institute of Canada (11919- 82 Street NW, Edmonton, Alberta, Canada, aamardon@yahoo.ca).

**Introduction:** The NASA Exploration Systems Architecture Study (ESAS) discharged in 2005 evaluated the top-level investigation transport and framework engineering and prerequisites: First, characterize top level necessities and designs for group and load dispatch frameworks. Second, create reference investigation design idea to help continued human and automated lunar investigation choices. Also, third, distinguish key advancements required to empower and improve reference investigation frameworks and re-prioritization of close term and far term innovation investments. Part of the examination researched the conveyance of forces to refuel NASA and business resources. This opens up conceivable outcomes for expanded business association and the potential for future Martian investigation battles. This can possibly change Mars transportation engineering by not expecting to put and execute in complex and cost-restrictive impetus frameworks, for example, SEMP and CP which are in numerous present ideas and plan for future missions. It is imperative to consider that transportation, design, and directions on Mars are required not just for future space investigation mission to Mars, yet in addition for the future settling and occupying of the planet by humankind. The future colonizers of Mars will most likely colonize the planet by living in the magma tubes on Mars' surface, in this manner guaranteeing their wellbeing and assurance. Yet, while considering the advantages of the colonization of Mars, we should likewise consider that there are different difficulties that must be confronted while possessing it. Some of which are: lethal degrees of radiation, presentation to rap-inactively changing outrageous temperatures, just as falling micrometeorites. Upon entry to Mars, the future colonizers of Mars, will be looked with this issues and subsequently, they should be considered and examines all the more profoundly so as to guarantee the future occupants of the planet, are free from any potential harm upon their landing to the planet. The possessing of blemishes, will help humankind from various perspectives, and along these lines tackle huge numbers of the issues that mankind is confronting today, for example, overpopulation. What's more, as the issue of overpopulation will be settled, the issues of creating enough sustenance, and having enough characteristic assets will be settled, in that capacity and will be under human control. Since, it must be taken into air conditioning tally that our normal assets are finishing, and as the total populace develops, more nourishment is required to encourage it, along these lines by occupying Mars, humankind will likewise tackle those issues, yet have the option to share in the rich common assets of Mars. Along these lines, the settling and possessing of the planet Mars is to be obviously analyzed and progressed in the direction of, if to be figured it out.

**Research:** An as of late proposed mission engineering structured by Beckman, Vogel, Peck and Patterson (2018) was intended to meet the NASA force resupply capacity subject. The significant structure segments incorporate the terminal for Martian and Extraterrestrial Transport Resupply (thus known as DeMETR) station alongside the High-circle Resupply Module for Exploratory Spacecraft (HRMESH). The previous will convey 19.6 T of Xenon and a joined 8.1 T of dinitrogen tetroxide (NTO) and Monomethylhydrazine (MMH) oxidizer to the Deep Space Transport in cis-lunar space. The mission structure would start with propelling the completely charge stacked DEMTRE station on the Space Launch System (SLS). Through a progression of complex steps and moving of drive, detachment, and changes in accordance with entering lunar circle utilizing a mind boggling appraisal of pre-decided direction structures, DeMETR will solidly go into lunar far off retrograde circle (LDRO). When this is finished, HRMES can likewise be propelled utilizing SLS and will pursue a similar direction as DeMTR to enter LDRO. A delineation of the mission structure is as per the following: This sort of in-space charge transfer, whenever demonstrated effective, would incredibly build the general moderateness and supportability of long haul Mars investigation crusades. To further survey the possibility and increase extra bits of knowledge of execution, prerequisites, coordinations and outcomes, these kinds of vehicles and framework at their different phases of plan, advancement and usage can be modeled to inexact execution utilizing a mix of similar to frameworks and reproductions. This may incorporate set up demonstrating instruments and methods found in POST for direction enhancement, APAS for streamlined features, NAFCOM for cost displaying and Monte Carlo examination for vulnerability analysis. For humankind to occupy the planet without being looked with the numerous issues of the planet Mars, for example, falling micrometeorites, presentation to extraordinary and quickly evolving temperatures, just as introduction to large amounts of radiation, is by possessing and living in the planet's magma tubes. While on Mars' surface, radiation levels are a lot higher than those on earth, and presentation to such lethal degrees of radiation is both unsafe to the human body, and even savage. Radiation comes from numerous points of view on Mars' surface, for example, sun oriented flares which are established comparatively to the sun powered breeze, yet the individual particles hold higher energies, and galactic inestimable beams which are made out of extremely high vitality particles, generally protons and electrons. Likewise, it is critical to consider that while on earth we have an environment and attractive field which can give adequately extraordinary security from the abnormal amounts of radiation, while Mars needs it. Presentation to extraordinary temperatures in like manner must be considered, for Mars is found further from the sun, than the earth, hence the temperatures on Mars are a lot colder than on earth. The normal daytime temperature on Mars in the winter season is about - 80 degrees Fahrenheit, or - 60 degrees Celsius in the daytime, while about - 195 degrees Fahrenheit, or - 125 degrees Celsius around evening time. In the mid year, the normal daytime temperature is warming up to around 70 degrees Fahrenheit and 20 degrees Celsius and the night normal temperature is about - 100 degrees Fahrenheit and - 73 degrees Celsius individually. Presentation to such extraordinary temperatures can make different sorts of mischief the human body, accordingly magma cylinders can give the important safe house, so as to endure such outrageous temperatures. Another issue to be represented is the issue of falling micrometeorites. Micrometeorites are little and unbelievably brisk falling bits of room flotsam and jetsam that can cause different effects on space explorers, contingent upon the size of the micrometeorite and the speed at which its voyaging. Despite the fact that most micrometeorites don't achieve earth's surface be-cause they vaporize by the significant measures of warmth that are brought about by the grating of going through earth's air, while in space there is no barometrical spread that would secure a shuttle or a space walker for a situation of falling micrometeorites. Those issues should genuinely be considered and innovation and transportation must be additionally improved and it must be guaranteed that the innovation is fit for shielding humankind from the issues they are to confront. In this way, enabling humankind to settle and occupy the planet Mars.

**Conclusion:** Different advances, and upgrades in transportation will be required for both future space investigation missions, just as the future settling and possessing of the planet Mars. Accordingly, it must be presumed that both future Martian missions which are basic for the future investigation of the planet, just as the future occupying of the planet, which would resolve a considerable lot of the issues we as of now face on earth, must be genuinely considered and profoundly contemplated, for they would bring an enormous advantage to mankind.

[4]

**References:** [1] National Aeronautics and Space Administration. (2005). NASA's Exploration Systems Architecture Study. Retrieved from https://www.nasa.gov/pdf/140649main_ESAS_full.pdf

[2] Beckman, E., Vogel, E., Peck, C., & Patterson, N. (2018). Depot for Martian and Extraterrestrial Transport Resupply (Master's thesis). Retrieved from https://pdfs.semanticscholar.org/eded/7018e5a2ef055fb0294087a1578bcd99e9c4.pdf

[3] Tanner, C., Young, J.J., Thompson R.W., & Wilhite, A.W. (2006). On-Orbit Propellant Re-supply Options for Mars Exploration Architectures. International Astronautical Congress, October 2005, Fukuoka, Japan.

[4] How far away is Mars? (n.d.). Retrieved from https://boingboing. net/2015/10/26/how-far-away-is-mars.html

**Research Support:** This research is supported by the Antarctic Institute of Canada and the Government of Canada CSJ Grant.

# Utilization of Resources on the Moon.

Gordon Zhou[1], Svetozar Zirnov[2], Austin Mardon[3], [1]The Antarctic Institute of Canada,(#103, 11919- 82 Street NW, Edmonton, Alberta, Canada, aamardon@yahoo.ca).

**Introduction:** Structures on the lunar surface will challenge contemporary thoughts of auxiliary examination by basic and structural architects, just as originators, constructors and coordination organizers. Uncovered home units will confront numerous issues with response to the extraordinary lunar temperature cycles and impacts of high vacuum. Uncovered material and basic exhaustion because of extraordinary lunar temperature cycles and temperature affect-ability differential on ceaseless auxiliary parts must be tended to night time lows of - 110°C (- 170°F) would mean originators must take a gander at the potential fragile cracks and stress focuses inside the potential material. A potential fractional arrangement is inside to pressurize supporting individuals, for example, those for clasping, hardening and propping to meet well being and unwavering quality prerequisite. The flighty idea of the lunar condition requires minimization of hazard to a worthy level. Stacking must consider the 1/6 gravity of the moon. This implies a structure will have multiple times the weight-beating limit (dead weight) on the moon as on the Earth. Notwithstanding, it can-not be expected that the structure can bolster more load because of this reality. This would possibly be valid if the material is straightly versatile. In any case, most materials have a non-direct. Current building thinking and configuration depends on the point of confinement state conditions. Chuaetal (1992) propose a nonlinear hyperbolic pressure strain model to all the more likely reflect how structure-regolith recreations should be possible utilizing the limited component approach. This is actually the reason clarifying the usage utilizing kg-power (estimations without gravity segment). Basic parts must display a degree of repetition as in statically vague structures. This infers burdens are redistributed to a balance state when individuals are to fall flat. A degree of worthy hazard and well being elements should be inferred. Since, the condition's on the moon's surface are very harsh, safety measures must also be taken into account. One way of making sure that an astronaut is safe while on a mission to the moon is through the use of the lunar lava tubes. One of the ways by which lava tubes may be used for future moon expeditions is by providing astronauts a shelter from the rough conditions on the moon's surface, which include falling micrometeorites, exposure to extreme temperatures, as well as fatal levels of radiation. Micrometeorites are small pieces of space debris which may have various impacts on the astronauts in the expedition, depending on the size and speed of the micrometeorites falling. Even though most of the falling micrometeorites do not reach earth's surface because they vaporize by the large amounts of heat that are generated by the friction of passing through earth's atmosphere, in space there is no atmospheric cover that would protect a spacecraft or a spacewalker in a case of falling micrometeorites. Also, as previously mentioned Another issue that must be taken into account is the issue of exposure to fast-changing extreme temperatures, for on the moon there is no atmosphere like on the earth and temperatures greatly vary between the day time highs and night time lows. Exposure to such extreme temperatures that are quickly-changing can cause various kinds of harm to the human body. Another major issue that should be taken into account is the exposure of astronauts to fatal levels of radiation while on the moon's surface, which may come in various ways such as: solar flares which are constituted similarly to the solar wind, but the individual particles hold higher energies, and also galactic cosmic rays which are composed of very high energy particles, mostly protons and electrons. Thus, just as astronauts face those various issues, while on missions to the moon, humanity will have to face similar issues when it will be inhabiting it, in the near future.

**Research:** Inflatables have for quite some time been proposed as a plausible and a financial strategy for a lasting lunar base. The inflatable pressurized tractable structure of fiber composites offers radiation protecting under local regolith and little temperature varieties. Erectable tetrahedral, hexahedral, octahedral structures have likewise been proposed and offers huge numbers of indistinguishable advantages from inflatables. The different geometrically arranged space outline components can be effectively expandable and rushes to build and introduce. As individuals from these structures are not locally discovered, it must be pre-manufactured and brought to the moon. At present it isn't financially achievable. It must be truly considered and taken into account that the utilization of lunar resources is essential to protect the health and well being of astronauts on a mission in space, as well as to ensure their safety at all times while on the mission. Lunar lava tubes can also act as a storage for various things, including medical supplies, food water, and even space equipment, thus keeping it in good condition, without damage. As the article in the argoverse website states: [4] " One other idea that has been proposed for the case of the Moon is that the sheltered environment and consistently cold temperatures in lava tubes may serve as a kind of trap for water ice and other volatiles (Billings, 1991, p. 256) ". Thus, helping astronauts survive during emergency situations. Thus, being a confined space it may also meet other astronaut's necessities, such as water, and dehydration. Since, confined spaces are cooler in temperature and are able to generate ice at times, the ice could be used either as water directly, or may be heated up using a piece of technology and turn into water that will be able to help the astronaut's in the mission, in case of a lack of water, and likewise help them to avoid dehydration. At times it may even save an astronaut's life, since water is the largest component of our bodies. This may also be used by the future human inhabitants as a source of water in a situation where water resources may become scarce. Thus, lunar lava tubes must be taken into account as a way of protecting astronauts and other humans in an emergency situation, or even as large storage facilities for storing food, medicine, or various kinds of equipment.

**Conclusion:** This paper shows a synopsis of squeezing issues encompassing the planning, engineering and development of lunar home units. Auxiliary honesty relies upon different flighty factors present on the moon. Temperature and regolith varieties must be considered into the plan standard of the structure. Essential, Material and Structural mechanics and conduct are reliant on these factors. Because of various factors inside the extent of planning of lunar structures, a strategy model for the thought of disappointment modes that vary from earthly structures must be made. The test during the structure stage is the powerlessness to test plan models under lunar condition. A sensible testing situation can be not practically tried. This thus does not enable architects and fashioners to successfully and precisely assess the total basic life cycle. The separation far from Earth related to surprising expenses related with vehicle of material to the lunar surface recommends the requirement for the utilization of local material. This is otherwise called In-situ Re-source Utilization (ISRU). This will be critical however future achievability investigation into this theme must be inquired about. Humanity's inhabiting of the moon is an important step in its history, that must be clearly planned and executed carefully. Since, astronauts are being effected by various kinds of issues such as, rapid temperature changes, exposure to high levels of radiation, as well as falling micrometeorites, humanity must be ready to face such issues. Lunar lava tubes may be used both by astronauts, for future space missions, but also by those humans who will be inhabiting the moon in the near future, as both emergency shelters, and as storages for food, medicine, and equipment, among other necessities.

[5]

**References:** [1] Benaroya, H., Hernold L, and Chua, K.M. (2002). Engineering, Design and Construction of Lunar Bases. Washington, DC. American Society of Civil Engineers.

[2] Miller, eds., ASCE, New York, 1952– 1963. Lunar Bases and Space Activities of the 21st Century. (1985). Lunar and Planetary Institute, p. 361-429.

[3] Chua, K. M., Johnson, S. W., and Sahu, R. (1992). Design of a support and foundation for a large lunar optical telescope. Engineering, construction and operations in space III, W. Z. Sadeh, S. Sture, and R. J. Miller, eds., ASCE, New York, 1952–1963.

[4] (n.d.). Retrieved from http://www.argoverse.com/LAVATUBE.html

[5] We Could Be Living On The Moon In 10 Years Or Less. (n.d.). Retrieved from https://www.popsci.com/we-could-be-living-on-moon-in-10-years-or-less/

**Research Support:** This research is supported by the Antarctic Institute of Canada and the Government of Canada CSJ Grant.

# Video Games for Astronauts in Space to Combat with Mental Illness.

James Fisher[1], Svetozar Zirnov[2], Austin Mardon[3], [1]The Antarctic Institute of Canada (11919- 82 Street NW, Edmonton, Alberta, Canada, aamardon@yahoo.ca).

**Introduction:** Video games may definitely be a fun and enjoyable way to mostly spend time, both as individuals, as well as with family and friends, but also they particularly are a way of combating mental illness for astronauts, while on a mission to space in a particularly major way. Each time astronauts literally exist the earth's atmosphere and definitely are going on a mission to space, there are always risks of them developing mental illness in a very major way. Astronauts on a space mission, often essentially find themselves working in an environment that for the most part is largely pressurized, as well as literally is very risky, which definitely shows that each time astronauts specifically exist the earth\'s atmosphere and particularly are going on a mission to space, there for the most part are always risks of them developing mental illness in a subtle way. The risk of astronauts developing various kinds of mental illnesses while on a mission, definitely is very high, and thus measures must for the most part be taken in order to basically assure that the astronauts going on the mission actually remain healthy both while on a mission, as well as when back from it, showing how the risk of astronauts developing various kinds of mental illnesses while on a mission, for the most part is very high, and thus measures must specifically be taken in order to really assure that the astronauts going on the mission actually remain healthy both while on a mission, as well as when back from it in a basically big way. One kind of common experience that definitely many astronauts definitely have while returning from a mission, particularly is that of hallucinations, showing how video games may basically be a fun and enjoyable way to for all intents and purposes spend time, both as individuals, as well as with family and friends, but also they basically are a way of combating mental illness for astronauts, while on a mission to space in a subtle way. Hallucinations literally come in basically many ways really such as, flashes of light appearing from nowhere, among others, very further showing how hallucinations generally come in particularly many ways generally such as, flashes of light appearing from nowhere, among others in a subtle way. In order to mostly reduce the risk of the development of mental illness in space, bringing video games into the space craft may for all intents and purposes be helpful, very contrary to popular belief. As astronauts will essentially be able to actually play video games in space, this will distract them from the reality around them when they basically are stressed, or for the most part are having hallucinations, demonstrating that as astronauts will essentially be able to mostly play video games in space, this will distract them from the reality around them when they generally are stressed, or literally are having hallucinations in a subtle way. Video games can really be set up in the space craft, by installing a TV, and a game console, and likewise bringing along a few games that may definitely be particularly played while on a space mission, so in order to basically reduce the risk of the development of mental illness in space, bringing video games into the space craft may basically be helpful, which for all intents and purposes is quite significant. It is really essential to note that since, the risk of the development of mental illness generally is actually such basically high in space, it must mostly be assured that astronauts definitely remain healthy at all times, both when on the mission, and when returning from it, demonstrating how in order to basically reduce the risk of the development of mental illness in space, bringing video games into the space craft may be kind of helpful.

**Research:** Two games I would actually play are: The Legends of Zelda: Breath of the particularly Wild and NHL 19 in a subtle way. In Breath of the basically Wild you could spend sort of large amounts of time, thus ensuring that the astronaut engaged in the video game, kind of has a very low risk of developing mental illness in a subtle way. If you kind of want a actually little sports game while in space you can basically play NHL 19 in a really big way. It's a kind of good game actually play some NHL hockey in space, this will specifically give the astronaut engaged in the video game something to mostly do while waiting in space for definitely your destination, which literally is fairly significant. While playing the game, it would literally make you mostly feel you for the most part are in an NHL game, so when an astronaut actually is on a mission, he/she could basically keep their focus on the game, rather than the reality around them, thus assuring that when the astronaut returns from the mission, he/she will not particularly be suffering mental illness, or so they basically thought. LOZBOTW literally has a lot things to really do in it, basically such as korok seeds side quest and shrines, kind of contrary to popular belief. There\'s lots of tasks in the game to perform, and if this really is not helping you could kind of try playing NHL 19, where it literally is a fast game there particularly are a fairly few modes in a actually big way. If you for the most part have an internet connection, you could basically choose to kind of proceed with the online modes or you could use the offline modes, very such as a generally general definitely match ,season, essentially be a pro, playoffs, shootouts, drafting players, and signing players, demonstrating how there's lots of tasks in the game to perform, and if this is not helping you could for all intents and purposes try playing NHL 19, where it mostly is a fast game there definitely are a very few modes, contrary to popular belief. There generally are lots of tasks to perform, which in particularly turn will for all intents and purposes help avoiding the development of mental illness, which for all intents and purposes is fairly significant. Thus, we must note that video games mostly are to kind of be considered definitely more seriously, since they for all intents and purposes are able to basically help essentially reduce the risk of the development of mental illness by astronauts in space, which for all intents and purposes shows that if you basically have an internet connection, you could essentially choose to specifically proceed with the online modes or you could use the offline modes, definitely such as a definitely general particularly match ,season, definitely be a pro, playoffs, shootouts, drafting players , and signing players, demonstrating how there\'s lots of tasks in the game to perform, and if this for all intents and purposes is not helping you could kind of try playing NHL 19, where it for the most part is a fast game there generally are a fairly few modes, which is quite significant.

**Conclusion:** It definitely is extremely fundamental to for the most part take note of that since, the danger of the advancement of psychological maladjustment and is generally high in space. It should for the most part for all intents and purposes be guaranteed that space explorers unquestionably particularly stay very solid consistently, both when on the mission, and when coming back from it, exhibiting how so as to fundamentally basically diminish the danger of the improvement of actually dysfunctional behavior in space, bringing computer games into the space specialty might basically be somewhat useful. This may provide astronauts with a solution of lowering the risk of mental illness development, while in space. Video games will essentially keep astronauts busy, while on a space mission, which will reduce the risk of mental illness development, not only while on the mission, but also when returning to earth. Thus, we are to conclude that video games are a means of helping astronauts reduce the risk of mental illness development while in space, as well as assuring that they will not develop it when returning to earth, and as such must be taken into account more seriously.

[2]

**References:** [1] Bell, V. (2014, October 05). Isolation and hallucinations: The mental health challenges faced by astronauts. Retrieved from https://www.theguardian.com/science/2014/oct/05/hallucinations-isolation-astronauts-mental-health-space-missions

[2] Funny cute astronaut playing video game with his best friend. (n.d.). Retrieved from https://www.123rf.com/photo_100523380_stock-vector-funny-cute-astronaut-playing-video-game-with-his-best-friend-on-their-own-space-design-for-printed-t.html

**Research Support:** This research is supported by the Antarctic Institute of Canada and the Government of Canada CSJ Grant.

# Evaluation of the Exclusion of Dental Services from Essential Medical Services during COVID-19

**Louis S. Park[*1,2], John Johnson [1,3], Peter Johnson [1,3], Jilene Malbeuf [1,3], Jasrita Singh [1,2], Austin A. Mardon, CM Ph.D.[1,3]**

[1] Antarctic Institute of Canada, Edmonton AB, Canada

[2] McMaster University, Faculty of Health Sciences, Hamilton ON, Canada

[3] University of Alberta, Edmonton AB, Canada

### Introduction

The COVID-19 pandemic placed many non-essential services on hold. However, the boundaries between essential and non-essential services are unclear in the medical sector, particularly in regards to dentistry. Essential services are daily services essential to preserving life, health, public safety and basic societal functioning. Hence, amid the COVID-19 pandemic, hospitals are triaging and treating patients normally, accepting a wide range of patients with varying degrees of emergency and illnesses that pose a risk on an individual's health and safety. However, bearing the same risk of transmission and equally preserving public health, most dental procedures have been suspended, and resorted to telemedicine. Dentists are not well protected in their work environments and people who are facing minor dental illnesses are left untreated indefinitely, increasing their risk of severe, dental illnesses in the future.

We discuss the guidelines of essential medical services set out in 2020 during the COVID-19 pandemic and implications and specific challenges that arise from the closure of 'non-essential' dental treatments.

### Which dental procedures are considered essential?

The Royal College of Dental Surgeons of Ontario (RCDSO) has only declared dental emergency procedures as essential, which constitutes trauma, infection, prolonged bleeding, and severe pain. [1] All other regular urgent procedures that would maintain a patient's oral health were halted indefinitely. Also, the RCDSO has further restricted dentistry through the use of teledentistry; a virtual provision of professional dental advice via communication technologies [2]—one not being readily used or recommended in hospitals. Some Canadians lack communication devices and most dental offices are not technologically prepared to offer teledentistry. This limits the number of patients they can manage and leads to the closure of dental clinics, placing personal financial consequences on dentists as well. In fact, about 78.7% of dental emergencies presented at hospitals are semi- to non-urgent and are diverted to local dental offices, leaving these patients without guaranteed treatment. 3 This would further affect COVID-19 response as it would be difficult to relieve emergency rooms, maintain low cost and high quality of care during a pandemic.

Furthermore, oral infections are known to affect the pathogenesis of systemic diseases including cardiovascular disease, pneumonia, and diabetes. 4 However, with indefinite halt in place, many dental patients are unable to prevent or detect time-sensitive dental illnesses like tooth decay, gum disease, oral cancer or vitamin deficiencies that would normally have been treated. This is especially problematic when considering the large population of frail and elderly individuals in Canada who are significantly associated with poor oral health. 4 Hence, the exclusion of dental care services poses a threat on the oral health of Canadians, thereby increasing risk of possible systematic diseases, poor quality of life, and if untreated, time-sensitive dental problems.

### What special COVID-19 preventative challenges arise in dental clinics?

COVID-19 is known to spread via bodily fluid particles such as mucus and saliva, substances dentists are connected with during dental procedures. Additional challenges arise from the nature of the procedures in which the clinician must be in great proximity, face-to-face with the patient, making it impossible for dentists to comply with the social distancing measures. In addition, with the scientific brief from the World Health Organization (WHO), which has suggested that COVID-19 may be capable of airborne transmission, 5 dentists are at an even higher risk for COVID-19. This may substantiate and justify the exclusion of dentists who are most vulnerable and at a greater risk of transmission of COVID-19 from essential medical services for their protection.

Nonetheless, while these safety measures and the closure of dental services maintain safety and protection from the coronavirus, this poses significant personal financial consequences on private offices and employees.

**Personal Protective Equipment supply and demand**

The College of Dental Hygienists of Ontario and RCDSO has established high-demanding, preventive guidelines and protocols. These protocols mandate the use of N95 masks, gowns, and other personal protective equipment (PPE) such as face shields and gloves, which demand a higher supply. As COVID-19 cases are increasing, it is crucial to allocate PPE to those in the hospitals interacting with active COVID-19 patients, especially during a period of shortage. With such high demands in place with an inadequate supply, the "essential" boundary becomes apparent. Clinicians interacting with active patients should rightfully be considered more "essential" and be given priority in regards to PPE.

However, according to the Priority setting of PPE released by the Ontario Provincial government, the Secondary Allocation Principle relies on first-come-first-serve or lottery, 7 which is not risk level- nor demand-focused. These principles limit dentists' allocation for PPE, threaten their safety, and it raises a concern of whether every dental office would even be able to provide emergency care.

**Conclusion**

While essential medical services such as emergency rooms directly treating COVID-19 patients should be prioritized, public health control measures pose challenges to dental services and increase the risk of poor oral health of Canadians amid indefinite lockdown measures. Safety protocols have been implemented in Canada to prioritize essential COVID-19-related healthcare, which caused a redeployment and reduction of dental clinicians. Essential medical care should be expanded to include more urgent dental procedures, not limited to virtual appointments nor to symptoms that classify as emergencies. The diverse implications must also be re-considered to ensure the well-being and health of citizens and clinicians, as well as an equitable and need-based distribution of PPE. The COVID-19 pandemic raises the question of whether the current measures on halting dental procedures must be re-evaluated for future pandemics.

### Acknowledgement

We thank and acknowledge the financial support of the Canada Service Corps, TakingITGlobal, and the Government of Canada.

### References

1. Definitions of emergency, urgent and non-essential care. *Royal College of Dental Surgeons of Ontario*. May 2020. PDF. https://az184419.vo.msecnd.net/rcdso/pdf/standards-of-practice/ RCDSO_COVID19_Definition s.pdf. Accessed August 4, 2020.

2. Jampani ND, Nutalapati R, Dontula BS, Boyapati R. Applications of teledentistry: A literature review and update. *J Int Soc Prev Community Dent*. 2011;1(2):37-44. doi:10.4103/2231-0762.97695

3. Wall T, Nasseh K, Vujicic M. Majority of dental-related emergency department visits lack urgency and can be diverted to dental offices. Health Policy Institute Research Brief. American Dental Association. August 2014.
    http://www.ada.org/~/media/ADA/Science%20and%20 Research/HPI/Files/HPIBrief_0814_1.as hx

4. Li X, Kolltveit KM, Tronstad L, Olsen I. Systemic diseases caused by oral infection. *Clin Microbiol Rev*. 2000;13(4):547-558. doi:10.1128/cmr.13.4.547-558.2000

5. Transmission of SARS-CoV-2: implications for infection prevention precautions. World Health Organization. https://www.who.int/news-room/commentaries/detail/transmission-of-sars-cov-2-implications-f or-infection-prevention-precautions. Published July 9, 2020. Accessed August 4, 2020.

6. Coulthard P. Dentistry and coronavirus (COVID-19) - moral decision-making. *British Dental Journal*. 2020;228(7):503-505. doi:10.1038/s41415-020-1482-1

7. Ontario Health. Priority Setting of Personal Protective Equipment – Within Health Care Institutions and Community Support Services. Published March 25, 2020. Accessed August 6, 2020. https://www.wrh.on.ca/uploads/Coronavirus/Ethics_Table_Policy_ Brief_3_PPE_Within_Health _Care_Institutions_Community_Support_ Services.pdf

# III

# Robert Heinlein

**BY: Svetozar Zirnov, Austin Mardon, Peter Johnson, John Johnson, and Elisia Snyder**

Known as the "dean of science fiction writers, Robert Heinlein was one of the most important science fiction writers of the 20th century. Robert Heinlein's short stories feature like a prophesy for the future of the United States, in which the United States goes through rapid changes and private enterprise sets itself up on the moon. Robert Heinlein wrote many adult science fiction books, as well as short story books. There was also a book beloved by Charles Manson, called "Stranger in a Strange Land", the book featured in itself many fictional religions. In his books, he also used to pass readers the idea that the unknowing are being controlled, or possessed by aliens. Robert Heinlein, together with Isaac Asimov, and Arthur C. Clarke, ushered in the "Golden Age of Science Fiction". Science fiction has gathered itself a fairly large audience in the 1940's, people fell in love with this type of writing. In 1974 Robert has become the Grand Master of the science fiction genre. Robert was born in the state of Missouri before his family decided to move to Kansas City, where most of his youth years were spent. Robert spent large amounts of his time in the library reading sci-fi and various fantasy books, by then popular authors, such as H.G. Wells, Edgar Rice Burroughs, and Olaf Stapledon. Robert in his lifetime has made many predictions, one of which was that interplanetary travel is real, and you are able to accomplish it, but in order to do so, you need to invest large amounts of money. This prediction is self evident, since today we have astronauts going to space, and conducting various missions in Mars, the Moon, and outer space. But, in order for astronauts to go to those missions, they are to invest large amounts of money. He clearly predicted that neither the mankind, nor our civilization will be destroyed. There prediction so far also holds to be true in that the human race is still here and thriving, our civilization is still evolving and is not destroying itself. Robert has predicted that by the end of the century the human race will explore the whole solar system and the first ship that will reach a star will become a building. Humanity has always fascinated itself with the exploration of space and had the intended goal of becoming an interplanetary species. Thus, it is quite possible that as the time passes technology will evolve and become more efficient and thus humanity will be able to explore more distant planets, as well as possibly colonizing them. Given the current human interest and intention, the exploration of our solar system is inevitable. Robert has predicted that intelligent life will be found on Mars. This prediction may find itself coming to be true in the near century. Given that NASA has got signals back from the planet, as well as there have been found basic life supporting resources, such as water, in the form of ice. Given the current human interest in the exploration of extraterrestrial life, the possibility of it being found on Mars is entirely possible. Robert has predicted that humanity will not achieve a "world state" in the near future, but the ideology of communism will vanish from this planet. This prediction may come true in the near future, since we can see many countries and republics that were formerly under communist rule, are turning towards either social democracy, in which there is a mixed market economy, or otherwise towards more conservative and neo-liberal capitalist systems. Robert has predicted that the main goal of worldwide physicists will be to learn to control the force of gravity. The prediction is unfolding right before our eyes given that physicists more and more often invoke the issue of controlling gravity, and the interest in the field is growing rapidly. Robert has claimed that it is a fact that if an attack came from space, there will be no way by which the authorities could repel it. Given the current situation, where countries mostly focus on defending themselves from the invasions of other countries, if the invasion would come from space, humanity will not be able to use it great weaponry and technology, to stop the invasion from happening. Robert has predicted that humanity around the world will be getting more and more hungry, and the world's food resources will become scarce. Given the current situation, we are facing the issue of the Covid-19 pandemic, where many people have lost their working positions, thus not earning enough money to be able to sustain themselves and their families. Also, as we are facing the issue of climate change, crops will also become either of poor quality, or will be reduced in number, since places that were previously hot and dry, are becoming cooler, and places that were cold, are becoming warmer, this may have a large impact on crops, and as crops become scarce, so will the world's food supply. Robert also has predicted that our house phones will be of small size, fitting in our pockets, and will be able to record messages, answer some inquiries, as well as transmit vision. This is a self fulfilling prophesy, since the smart phones we own today, are quite similar to the phones being described by Robert, having the ability to record messages, answer some inquiries, by using Siri, and transmitting vision, through the various apps, such as Skype, Facetime, and Zoom. Also, those phones literally fit in our pockets. Those are just some of Robert's predictions, out of many others that he has made and are already coming true. Robert was not only a science-fiction writer, but also a genius and almost a prophet, since his predictions are truly fulfilling.

References:

1. Robert A. Heinlein. (2015, June 11). Retrieved July 16, 2020, from https://www.bookseriesinorder.com/robert-a-heinlein/

2. Https://www.fantasticfiction.com, W. (n.d.). Robert Heinlein. Retrieved July 16, 2020, from https://www.fantasticfiction.com/h/robert-heinlein/

3. 19 Predictions For The Future From Robert A. Heinlein. (2019, November 30). Retrieved July 16, 2020, from https://www.writerswrite.co.za/robert-a-heinleins-19-predictions-for-the-future/

# Robert Heinlein and American Wars

BY: Svetozar Zirnov, Austin Mardon, Peter Johnson, John Johnson, and Elisia Snyder

Robert Heinlein, the great science fiction writer, and the greatest of his kind, has predicted that many events some of which were political, economic, social, and scientific. Robert has been dubbed to be the one who ushered in "The golden age of science fiction". Many of Robert's predictions came true, and others that did not are starting to come true. One of Robert's great predictions was that the US will never start a war that can be prevented, but rather will fight those wars, that are needed to be fought to defend either its own territory, or the territory of its allies. The US has always had many allies, as well as, many enemies. Some of the US allies, would be Israel, Saudi Arabia, the UK, Australia, New Zealand, and Canada among many others. Among the enemies of the US, would largely be North Korea, Russia, and Iran, among others. The US has generally never started a war, even though it is the 3rd largest country by land in the world, as well as, it's a superpower. The US has a lot of weaponry and a lot of ways to defend itself, as well as, its allies during a war waged on either its own land, or the land of its allies. The US has usually also been a trustworthy partner, to those who considered making the US their ally. The US is either the first or second country in the world having such massive amounts of weapons, and other military supplies. Both the US and Russia also own a vast amount of nuclear weapons, which include nuclear rockets and bombs. Even though, pretty recently we were witnesses of heavy escalation between the US, Russia, and North Korea, the situation has largely calmed down, and is hopefully moving towards the better. The country of North Korea, has always been seen as a closed country, only those individuals residing in it are allowed to live there, those who are from outside the country, are to go back to their countries, as soon as their trip ends. The country has been ruled by iron and fist by its president Kim Jong Un. The currently carries a large array of weaponry, including nuclear weaponry. Even though many would call the country's system "Communist", in fact in 1991, the country has adopted a new ideology called "Juche", which resembled the rule of Joseph Stalin in the early Soviet Union. The ideology was most likely seen as a form of "Red Fascism". This is a remaking of the right wing ideology, in this sense, it is not the corporation who exploit their workers, like under the far-right version of the ideology, but rather, it is the state the exploits its working class. Red Fascism is seen as a separate ideology to Communism, and thus the two don't match. The Russian Federation, previously called "The Soviet Union", has been ruled by iron and fist by its current president Vladimir Putin. Putin's ideology is hard to examine, since while he takes more conservative views on social issues, economically he takes more moderate/liberal views. President Putin has been considered a moderate/centrist by Russian standards, while by some other standards he is considered more of a Fascist, having far-right views. Russia and the US have always been in conflicting relations, even since the time of the existence of the Soviet Union. The US has always been there going for the various peace missions, such as the NATO and UN missions. NATO and the UN (United Nations) have always been those organizations who have assisted in the various peace keeping missions around the world. They exist in order to assure that countries that are in conflict, or in a civil war, those organizations have always been there to assist those countries, in order to assure that all conflicts are being resolved in a peaceful and timely manner. The US has been involved in many civil wars, as well in the middle east. Conflicts, such as those in Syria, Egypt, Libya, among others. The US has been always the defender of peace around the world. There was not a time when the US has started a war by its self. The US has contributed a lot of their support and resources in various humanitarian missions around the world. The US has been a big help in the war in Syria, of which we just heard recently. The conflict is quite large, and it did not end quite yet. All peace missions conducted by the US, have contributed a lot for both resolving the conflicts that those countries had, as well as, the refugee camps which have helped many people survive. In the various refugee camps in Africa and the middle east, many people are able to get a meal and a drink. Those are things which the people residing in those countries, were not able to have enough of. The US has contributed a lot to save the vast amount of refugees from various countries in the world. The US has become a country of not only refuge, but also a country of prospects, for the poor and starving population. The US has also contributed a lot in the fight of World War Two. The US has done anything possible to end the war, and to end the continuous murdering and slaughtering of people. The second world war has caused many losses, among the many various nations around the world, and the ones that suffered the most, were the Jews. That same nation that has always been seen by the many as a target, a nation, that could not find itself a safe home, until the establishment of the government of Israel, has been affected the most, by the Nazi dictator Adolf Hitler, who sought to kill all Jews. The US is the country that has always stood up to the world's evil, and has done all possible to assure that peace is a human right in the world's various countries. As Robert Heinlein has predicted, the US does not start a war of itself, but rather does everything possible to assure that peace is maintained in this harsh time for the world.

### References:

1. The Moon Is A Harsh Mistress : Robert Heinlein : Free Download, Borrow, and Streaming. (1966, January 01). Retrieved July 21, 2020, from https://archive.org/details/TheMoonIsAHarshMistress_201701

2. Robert Heinlein, Moon is a Harsh Mistress: Essay Samples Blog. (2019, July 12). Retrieved July 21, 2020, from https://www.paperwritings.com/free-examples/robert-heinlein-moon-is-a-harsh-mistress.html

3. (n.d.). Retrieved July 21, 2020, from https://www.sparknotes.com/lit/The-Moon-Is-a-Harsh-Mistress/

# Robert Heinlein and Covid-19

BY: Svetozar Zirnov, Austin Mardon, Peter Johnson, John Johnson, and Elisia Snyder

Covid-19, the great pandemic of our time is seen as being one of the worst in human history. The pandemic has just begun, and then very rapidly brought the whole world under its influence. The world's largest countries, as well as the smallest have become victims of this deadly disease, and the world went into a deep quarantine. Countries are fighting with the virus in various ways and wanted to try to assure that their citizens are safe and are staying alive and healthy. The virus is very dangerous and has caused many people's deaths. Robert Heinlein, the great science fiction writer, who has ushered in the "the golden age of science fiction", wrote a book named :The Moon is a Harsh Mistress". In the book he describes a time in the year 2076 when humanity will be residing on the moon. The moon will become a colony for people who are unwanted here on earth. In order for them to survive in this harsh and unforgiving environment, the people exiled have created themselves a libertarian society. Their motto is TANSTAAFL, which stand for "there ain't such thing as a free lunch". There on the moon there is an authority called Luna, which governs the moon, as well as it is dealing with earth. This earth administration is importing all the necessities that are needed for Loonies to survive and exporting grain to earth, in order to assure that humanity does not face starvation. In order to stop the Luna administration for overusing the planet's resources, as well as avoiding an environmental disaster, the inhabitants of Luna wanted to rebel against their authority. The population organizes itself into a revolution, even though Luna's population is sure that they will not be able to conquer the forces of the Luna administration. Luna's inhabitants are faced with a great dilemma, they want to understand how their society came to be, and how does it function until today. The population concentrated on the political, as well as economic issues of a free society, and how they were operating. The inhabitants came to realize that government is a problem of itself and rebelled against it, in order to eliminate it. While living on the moon, its inhabitants came to a point where they had to deal with a deadly disease. While dealing with the disease, the inhabitants if the moon had to assure that they are keeping distance between themselves in order to assure that the disease is not going to spread around, and is not going to infect others. The disease was very contagious, and people had to stay indoors for large amounts of time, in order to not transmit the disease. This scenario is like a prophesy of our time, it clearly predicts Covid-19, its spread, and its effects on society in large. We see businesses closing down, in order to avoid employees getting sick with Covid-19. Schools are shutting down, and are being transferred to online studies, in order to assure that students remain well and healthy. Many people are trying to spend more time indoors, in order to avoid excess human contact, and to avoid the spread of Covid-19. Hospital workers are working on full capacity, in order to save live, many who are unable to keep up with the pace, are quitting their jobs. Just as Robert has predicted, we are facing a contagious disease, that is easily spreading around. Also, while describing the disease, Robert is also describing the effects that the libertarian ideology and politics have on society. Given the situation with the inhabitants on the moon, the libertarian ideology was not pleasing people, and likewise did not represent the people, that is why people rose up in order to rebel against the government. In today's words, and especially during the pandemic, we see many political parties around the world, adopting various forms of libertarianism. While libertarianism sounds good to many people, and the inhabitants of a certain country believe that they are given more rights and freedoms, this is entirely not true. The freedom that libertarianism offers, is the freedom of large businesses and corporations to purchase and privatize previously government owned commodities. Those commodities may vary, usually commodities such as health care, natural resources, education, beaches, parks, and other industries are being privatized and transferred from the ownership of the state to the ownership of large businesses and corporations. The exiles on the moon began to understand the way that the government was heading, thus they rebelled in order to assure that their rights and freedoms are being respected, and not the rights and freedoms of large businesses and corporations. Robert also points out that the people rebelled against the government and were seeking to eliminate it, thus creating the notion, that the moon's inhabitants wanted to have a form of anarchy. In an anarchist system, the government is usually repealed and humanity governs itself. Thus, there are no higher authorities, such as the government that tells you what should be done, and how it should be done. Anarchy promises a society where the rights and freedoms of each person are being respected, and each person is responsible for his/her own doing, thus people govern, or police themselves. In order for such a society to come into reality humanity is to learn how to self govern itself, and know its own measures. During the pandemic, we have learned that government intervention is essential for human survival, countries that have set up government support programs, have done way better than the countries that did not. Countries where the government has taken a more libertarian view ended up creating either way too few government support programs, or not creating programs at all. In those countries, people pretty much had to survive on the limited resources they had, even if they had too few of them. Robert Heinlein in his book "The Moon is a harsh mistress", has not only predicted the Covid-19 pandemic, but has also taught us a lesson, that government intervention in the economy is essential to both tackle pandemics, as well as for human survival.

**References:**

1. The Moon Is A Harsh Mistress : Robert Heinlein : Free Download, Borrow, and Streaming. (1966, January 01). Retrieved July 21, 2020, from https://archive.org/details/TheMoonIsAHarshMistress_201701

2. Robert Heinlein, Moon is a Harsh Mistress: Essay Samples Blog. (2019, July 12). Retrieved July 21, 2020, from https://www.paperwritings.com/free-examples/robert-heinlein-moon-is-a-harsh-mistress.html

3. (n.d.). Retrieved July 21, 2020, from https://www.sparknotes.com/lit/The-Moon-Is-a-Harsh-Mistress/

# Robert Heinlein and Freedom

**BY: Svetozar Zirnov, Austin Mardon, Elisia Snyder, Peter Johnson, and John Johnson**

Robert Heinlein, the incredible sci-fi author, and the best of his sort, has anticipated that numerous occasions some of which were political, financial, social, and logical. Robert has been named to be the person who introduced "The brilliant period of sci-fi". A considerable lot of Robert's forecasts worked out as expected, and others that didn't are beginning to materialize. One of the incredible expectations that Robert Heinlein has made is that freedom will slowly be removed from individuals over a period of time. The belief system of Libertarianism has been set up so as to underline individuals' privileges and opportunities. The belief system advocates a type of self-administration where people living in the general public have more opportunity to police themselves. While Libertarianism battles for human rights, it by and by doesn't go similar to the belief system of Anarchy, where the administration is as a rule completely nullified. Libertarianism may come in various structures, there are Libertarians who are more traditionalist and bolster free market Capitalism, and typically recognize as being on the conservative of the political range (the most well-known sort), and there are Libertarians who distinguish as being Socialist Libertarians, who bolster a going full bore Socialist society, just without the state possessing all the methods for creation. Those Libertarians as a rule distinguish themselves with the left wing of the political range. In the book he portrays a period in the year 2076 when mankind will live on the moon. The moon will turn into a state for individuals who are undesirable here on earth. With the goal for them to get by in this cruel and unforgiving condition, the individuals ousted have made themselves a libertarian culture. Their witticism is TANSTAAFL, which mean "there ain't such thing as a free lunch". There on the moon there is an authority called Luna, which administers the moon, just as it is managing earth. This world organization is bringing in all the necessities that are required for Loonies to endure and sending out grain to earth, so as to guarantee that humankind doesn't confront starvation. So as to stop the Luna organization for abusing the planet's assets, just as maintaining a strategic distance from a natural fiasco, the occupants of Luna needed to defy their power. The populace sorts out itself into a transformation, despite the fact that Luna's populace is certain that they won't have the option to vanquish the powers of the Luna organization. Luna's occupants are confronted with an incredible issue, they need to see how their general public became, and how can it work until today. The populace focused on the political, just as financial issues of a free society, and how they were working. The occupants came to understand that administration is an issue of itself and opposed it, so as to kill it. Despite the fact that left wing Libertarians exist, the most well-known sort is the traditional Libertarians, and the general public that Robert is depicting in his book would recognize all things considered. In his book, Robert is giving us how a Libertarian culture, while its battling for the rights and opportunities of its populace, simultaneously, is supporting privatization of fundamental products, that were once viewed as open. At the point when the administration evacuates its control over specific businesses, we are to comprehend that those enterprises are getting open for privatization, so the less the state has control over a specific industry, the higher is the opportunity for that industry to be privatized. Along these lines, in a Libertarian culture, we will see exceptional privatizations of items that were recently viewed as open, and state possessed. Fundamental products, for example, social insurance, parks, sea shores, public washrooms, the vitality part, and others, are getting open for privatization. Such a framework may cause its populace more damage than benefits. The populace might be denied of fundamental rights which it had been utilizing for quite a long time, the most basic of which might be social insurance. General medicinal services framework, for example, found in specific nations around the globe, including the Scandinavian nations, and Canada, has helped a great many individuals to save money on human services costs. In the event that it was not for the state possessed medicinal services framework, numerous individuals wouldn't have the option to bear to pay for their own human services. Such is the circumstance in the US, where the human services framework is private, and isn't possessed by the state. People living in the US, are paying a lot of cash to simply have the option to get social insurance administrations which are basic for their very own endurance. It ought not be like this, since social insurance is a human right, of which a human ought not be denied. A Libertarian culture, additionally shares numerous components with a Fascist one, then again, actually a Libertarian culture doesn't take things to such a degree, as to have a tyrant, or tyrannical standard. Both Libertarianism and Fascism, for the most part, are centered around giving more opportunity to huge business, and undertaking. This permits large business, to buy private companies, so as to evacuate pointless rivalry. The littler the measure of private ventures accessible, the higher turns into the riches that is being aggregated by the huge business. In the US, since the human services framework is private, the enormous pharma is making a lot of cash off of individuals' wellbeing, in this manner aggregating riches. In a Libertarian culture, anybody can what he/she wishes, hence huge business can likewise aggregate as much riches as possible, without the need to appropriate that riches to the nations' populace, since in a Libertarian culture, government intercession is restricted. A few nations around the globe, just as, numerous ideological groups have incorporated certain components of Libertarianism into their own constituent stages. Gatherings, for example, the Conservative Party of Canada, the US Republican Party, the Alberta popular United Conservative Party, among numerous others, have consolidated a few components of Libertarianism into their appointive stages. Libertarianism, is what is normally called a "twofold edged blade", since while it bolsters individuals in battling for their human rights and opportunities, it likewise is helping huge business to have their privileges and opportunities spoke to and regarded, so in this manner, while individuals have more rights and opportunities, so do the transnational enterprises, who are attempting to make benefit off items that used to be open, and a portion of those products might be basic administrations, without which, human endurance might be solidified. Since the belief system of Libertarianism is getting famous around the globe, it is turning out to be entirely conceivable that, as Robert anticipated, we will be living in a Libertarian culture in the coming occasions.

**References:**

1. The Moon Is A Harsh Mistress : Robert Heinlein : Free Download, Borrow, and Streaming. (1966, January 01). Retrieved July 21, 2020, from https://archive.org/details/TheMoonIsAHarshMistress_201701

2. Robert Heinlein, Moon is a Harsh Mistress: Essay Samples Blog. (2019, July 12). Retrieved July 21, 2020, from https://www.paperwritings.com/free-examples/robert-heinlein-moon-is-a-harsh-mistress.html

3. (n.d.). Retrieved July 21, 2020, from https://www.sparknotes.com/lit/The-Moon-Is-a-Harsh-Mistress/

# Robert Heinlein and Libertarianism

BY: Svetozar Zirnov, Austin Mardon, Peter Johnson,
John Johnson, and Elisia Snyder

Robert Heinlein, the great science fiction writer, and the greatest of his kind, has predicted that many events some of which were political, economic, social, and scientific. Robert has been dubbed to be the one who ushered in "The golden age of science fiction". Many of Robert's predictions came true, and others that did not are starting to come true. One of the great predictions that Robert Heinlein has made is that not only will humanity be residing on the moon, but also that they will have a full swing Libertarian society. The ideology of Libertarianism has been put in place in order to emphasize people's rights and freedoms. The ideology advocates a form of self-governance where individuals living in the society have more freedom to police themselves. While Libertarianism does fight for human rights, it nevertheless does not go as far as the ideology of Anarchy, where the government is being fully abolished. Libertarianism may come in different forms, there are Libertarians who are more conservative and support free market Capitalism, and usually identify as being on the right wing of the political spectrum (the most common type), and there are Libertarians who identify as being Socialist Libertarians, who support a full swing Socialist society, just without the state owning all the means of production. Those Libertarians usually identify themselves with the left wing of the political spectrum. In the book he describes a time in the year 2076 when humanity will be residing on the moon. The moon will become a colony for people who are unwanted here on earth. In order for them to survive in this harsh and unforgiving environment, the people exiled have created themselves a libertarian society. Their motto is TANSTAAFL, which stand for "there ain't such thing as a free lunch". There on the moon there is an authority called Luna, which governs the moon, as well as it is dealing with earth. This earth administration is importing all the necessities that are needed for Loonies to survive and exporting grain to earth, in order to assure that humanity does not face starvation. In order to stop the Luna administration for overusing the planet's resources, as well as avoiding an environmental disaster, the inhabitants of Luna wanted to rebel against their authority. The population organizes itself into a revolution, even though Luna's population is sure that they will not be able to conquer the forces of the Luna administration. Luna's inhabitants are faced with a great dilemma, they want to understand how their society came to be, and how does it function until today. The population concentrated on the political, as well as economic issues of a free society, and how they were operating. The inhabitants came to realize that government is a problem of itself and rebelled against it, in order to eliminate it. Even though left wing Libertarians exist, the most common type is the right wing Libertarians, and the society that Robert is describing in his book would identify as such. In his book, Robert is showing us how a Libertarian society, while its fighting for the rights and freedoms of its population, at the same time, is supporting privatization of essential commodities, that were once considered public. When the government removes its power over certain industries, we are to understand that those industries are becoming open for privatization, so the less the state has power over a certain industry, the higher is the chance for that industry to be privatized. So, in a Libertarian society, we will see drastic privatizations of commodities that were previously considered public, and state owned. Essential commodities, such as health care, parks, beaches, public washrooms, the energy sector, and others, are becoming open for privatization. Such a system may cause its population more harm than benefits. The population may be deprived of essential services which it had been using for centuries, the most essential of which may be health care. Public health care system, such as found in certain countries around the world, including the Scandinavian countries, and Canada, has helped millions of people to save on health care costs. If it was not for the state owned health care system, many people wouldn't be able to afford to pay for their own health care. Such is the situation in the US, where the health care system is private, and is not owned by the state. Individuals living in the US, are paying large amounts of money to just be able to receive health care services which are essential for their own personal survival. It should not be this way, since health care is a human right, of which a human should not be deprived. A Libertarian society, also shares many elements with a Fascist one, except that a Libertarian society does not take things to such an extent, as to have an authoritarian, or dictatorial rule. Both Libertarianism and Fascism, generally, are focused on giving more freedom to big business, and enterprise. This allows big business, to purchase small businesses, in order to remove unnecessary competition. The smaller the amount of small businesses available, the higher becomes the wealth that is being accumulated by the big business. In the US, since the health care system is private, the big pharma is making large amounts of money off of people's health, thus accumulating wealth. In a Libertarian society, anyone can what he/she wishes, thus big business can also accumulate as much wealth as it possibly can, without the need to distribute that wealth to the countries' population, since in a Libertarian society, government intervention is limited. Some countries around the world, as well as, many political parties have included certain elements of Libertarianism into their own electoral platforms. Parties such as, the Conservative Party of Canada, the US Republican party, the Alberta provincial United Conservative Party, among many others, have incorporated some elements of Libertarianism into their electoral platforms. Libertarianism, is what is commonly called a "double-edged sword", since while it supports people in fighting for their human rights and freedoms, it also is helping big business to have their rights and freedoms represented and respected, so thus, while people have more rights and freedoms, so do the transnational corporations, who are trying to make profit off commodities that used to be public, and some of those commodities may be essential services, without which, human survival may be hardened. Since the ideology of Libertarianism is becoming popular around the world, it is becoming very possible that, as Robert predicted, we will be living in a Libertarian society in the coming times.

**References:**

1. The Moon Is A Harsh Mistress : Robert Heinlein : Free Download, Borrow, and Streaming. (1966, January 01). Retrieved July 21, 2020, from https://archive.org/details/TheMoonIsAHarshMistress_201701

2. Robert Heinlein, Moon is a Harsh Mistress: Essay Samples Blog. (2019, July 12). Retrieved July 21, 2020, from https://www.paperwritings.com/free-examples/robert-heinlein-moon-is-a-harsh-mistress.html

3. (n.d.). Retrieved July 21, 2020, from https://www.sparknotes.com/lit/The-Moon-Is-a-Harsh-Mistress/

# Robert Heinlein and Life on Other Planets

BY: Svetozar Zirnov, Austin Mardon, Peter Johnson, John Johnson, and Elisia Snyder

Robert Heinlein, the great science fiction writer, has predicted many events, some of his predictions have come to pass, and some are still on their way, waiting their time to come to pass. Robert Heinlein was one of the greatest science fiction writers of his century. Robert has ushered in, what is now called "The golden age of science fiction". Robert has predicted many events, some were political, some scientific, some economical, and so far, many of his predictions have come to pass. Robert has also predicted that intelligent life will be found in outer space, specifically on Mars. Humanity for centuries, and even possibly from its inception have been focusing on the issue of whether intelligent life exists somewhere else in the universe, beside earth. That issue has been studied by many scientist, who have developed certain kinds of technology in order to specifically deal with the issue of whether there is life in outer space. History here on earth indicates that presence of visitors from outer space has been here on earth. Throughout the history of the development of the human civilization, the history of the civilization presents us with the story of how extraterrestrial beings that came from other planets from outer space have intertwined with our civilization. The oldest civilization that we know of as humans is the Sumerian civilization. The Sumerian civilization has left after itself many writings and written works regarding their history and origin. The Sumerians believed that there was a civilization called "Annunaki", or "those who from heaven to earth came". The Sumerians regarded them as gods and worshiped them as part of their religious rituals. The Sumerians clearly believed that it was this Annunaki civilization that has given humanity all the knowledge that it currently possesses. The knowledge they gave was the arts of farming, painting, singing, dancing, building, magic, technology, magic, among many others. Most of what our modern day civilization is based on, according to the Sumerians, was given to us by the Annunaki civilization. The Sumerians believed that the Annunaki civilization came to earth from a planet called Nibiru. The planet Nibiru has been the home planet of the Annunaki race. The Annunaki lacked many precious metals and minerals on their own planet, and they needed them in order to survive, so the Annunaki decided to come to earth in order to mine for those metals and minerals. It is important to note that in history we can see at the beginning only the "primordial man", those were the popular caveman that we knew of. While, all of a sudden we see human beings being able to farm, sing, dance, mine, paint, build, do magic, and create technology. If the Sumerian writings are to be trusted than all the knowledge that humanity currently possesses was given to us by this Annunaki civilization who came to earth in order to mine for precious metals and minerals, in order to save their planet from extinction. If we are to compare the various stories and history accounts given to us in the Sinai codex, that is the oldest version of the Holy Bible that was found, we can clearly see the similarities that those accounts have with the Sumerian writings, and they are so identical that they were as if copied and pasted into the Biblical account. This story helps us to learn more about our human history, as well as, to learn about the presence of an extraterrestrial races that were present here on earth. Most of the world religions available today, be it Judaism, Christianity, Islam, Buddhism, and Hinduism, among others describe in a simplified way the presence of extraterrestrial races here on earth. All those religions directly, or indirectly tell of the story of how an extraterrestrial race came to earth, and has contributed a lot to the moral, social, as well as, religious development of the human civilization. Archeology around the world also makes us pause and think again about our origins as humans. Archeology presents many facts regarding extraterrestrial races visiting earth in ancient times. Archeologists find corpses that are way larger than human, they find artifacts and pictures that clearly depict beings similar to giants co-existing with humans. As we dig more into history in order to better understand our past as humans, we are coming across all this large amounts of information, that helps us to realize that we already have been visited by extraterrestrial races here on earth. For a very long time, possibly even from humanity's conception as a species, humanity has been fascinated with the stars and planets. Humanity was always seeking the answer to the question of whether life exists in outer space. Humanity has established special forms of technology to try and contact other extraterrestrial races in outer space. NASA, the organization in the US that has been very curious in the issue, has got some signals coming from outer space, but yet it is not determined whether those signals came from an extraterrestrial race in outer space or not. Given that most life supporting supplies have been found on various planets in outer space, may indicate that extraterrestrial life is a reality. As humanity is progressing in its development of technology, with the passing of time humanity will be able to have more powerful technology that may assist humanity in its task of determining whether extraterrestrial life is possible in outer space. If extraterrestrial life is to be found in outer space, this may lead us to realize that we are not the only species in space, and thus we will have a different feeling of our existence. If ancient writings from ancient civilization, and ancient artifacts are to be trusted then there is a possibility of life in outer space, but humanity is just not in the point of time, or is not ready to communicate with them. As Robert Heinlein has clearly predicted extraterrestrial life will be found on Mars, and if humanity will be able to generate the correct technology, this prediction will certainly come true, it's only a matter o time.

References:

1. Robert A. Heinlein. (2015, June 11). Retrieved July 16, 2020, from https://www.bookseriesinorder.com/robert-a-heinlein/

2. Https://www.fantasticfiction.com, W. (n.d.). Robert Heinlein. Retrieved July 16, 2020, from https://www.fantasticfiction.com/h/robert-heinlein/

3. 19 Predictions For The Future From Robert A. Heinlein. (2019, November 30). Retrieved July 16, 2020, from https://www.writerswrite.co.za/robert-a-heinleins-19-predictions-for-the-future/

# Robert Heinlein and Outer Space

BY: Svetozar Zirnov, Austin Mardon, Peter Johnson, John Johnson, and Elisia Snyder

Robert Heinlein, the great science fiction writer, possibly the greatest of all time, it was him who has ushered in what is being dubbed as "the golden age of science fiction". Robert has written many science fiction books on various topics. While writing the various books, Robert has made many predictions regarding the future, some of which came to be true, while some are on their path to realizing themselves. Robert has made many predictions which include, but are not limited to, economics, poverty, space travel, homelessness crisis, and the finding life on other planets. Throughout history humanity has been searching for evidence of the possibility of life on other planets. Robert has additionally anticipated that clever life will be found in space, explicitly on Mars. Humankind for quite a long time, and even conceivably from its origin have been concentrating on the issue of whether astute life exists elsewhere known to mankind, next to earth. That issue has been concentrated by numerous researcher, who have built up particular sorts of innovation so as to explicitly manage the issue of whether there is life in space. History here on earth demonstrates that nearness of guests from space has been here on earth. Since the commencement of the improvement of the human development, the historical backdrop of the progress presents us with the tale of how extraterrestrial creatures that originated from different planets from space have entwined with our development. The most established progress that we are aware of as people is the Sumerian human advancement. The Sumerian progress has left after itself numerous compositions and composed works with respect to their history and cause. The Sumerians accepted that there was a human advancement called "Annunaki", or "the individuals who from paradise to earth came". The Sumerians viewed them as divine beings and adored them as a feature of their strict ceremonies. The Sumerians unmistakably accepted that it was this Annunaki progress that has given mankind all the information that it as of now has. The information they gave was expressions of the human experience of cultivating, painting, singing, moving, building, enchantment, innovation, enchantment, among numerous others. The vast majority of what our cutting edge development depends on, as indicated by the

Sumerians, was given to us by the Annunaki human advancement. The Sumerians accepted that the Annunaki human progress came to earth from a planet called Nibiru. The planet Nibiru has been the home planet of the Annunaki race. The Annunaki needed numerous valuable metals and minerals on their own planet, and they required them so as to endure, so the Annunaki chose to come to earth so as to dig for those metals and minerals. It is critical to take note of that in history we can see toward the starting just the "early stage man", those were the famous mountain man that we was aware of. While, out of nowhere we see individuals having the option to cultivate, sing, move, mine, paint, assemble, do enchantment, and make innovation. In the event that the Sumerian works are trustworthy than all the information that mankind at present has was given to us by this Annunaki progress who came to earth so as to dig for valuable metals and minerals, so as to spare their planet from elimination. In the event that we are to think about the different stories and history accounts given to us in the Sinai codex, that is the most seasoned variant of the Holy Bible that was discovered, we can obviously observe the similitude's that those records have with the Sumerian works, and they are indistinguishable from such an extent that they were as though reordered into the Biblical record. This story encourages us to get familiar with our mankind's history, just as, to find out about the nearness of an extraterrestrial races that were available here on earth. The greater part of the world religions accessible today, be it Judaism, Christianity, Islam, Buddhism, and Hinduism, among others portray in an improved way the nearness of extraterrestrial races here on earth. Each one of those religions straightforwardly, or in a roundabout way recount the narrative of how an extraterrestrial race came to earth, and has contributed a great deal to the ethical, social, just as, strict improvement of the human development. Antiquarianism around the globe likewise makes us interruption and reconsider our beginnings as people. Pale history presents numerous realities with respect to extraterrestrial races visiting earth in old occasions. Archeologists discover cadavers that are route bigger than human, they discover curios and pictures that obviously portray creatures like mammoths existing together with people. As we delve more into history so as to all the more likely comprehend our past as people, we are going over this a lot of data, that causes us to understand that we as of now have been visited by extraterrestrial races here on earth. For quite a while, conceivably even from humankind's origination as an animal varieties, mankind has been interested with the stars and planets. Mankind was continually looking for the response to the topic of whether life exists in space. Humankind has built up extraordinary types of innovation to attempt to contact other extraterrestrial races in space. NASA, the association in the US that has been interested in the issue, has got a few signs originating from space, however yet it isn't resolved whether those signs originated from an extraterrestrial race in space or not. Given that most life supporting supplies have been found on different planets in space, may show that extraterrestrial life is a reality. As humankind is advancing in its advancement of innovation, with the progression of time mankind will have the option to have all the more remarkable innovation that may help humankind in its assignment of deciding if extraterrestrial life is conceivable in space. In the event that extraterrestrial life is to be found in space, this may lead us to understand that we are by all account not the only species in space, and along these lines we will have an alternate sentiment of our reality. On the off chance that old works from antiquated progress, and old ancient rarities are reliable then there is a chance of life in space, yet humankind is only not in the purpose of time, or isn't prepared to speak with them. As Robert Heinlein has plainly anticipated extraterrestrial life will be found on Mars, and if humankind will have the option to produce the right innovation, this expectation will absolutely work out as expected, it won't be long. Robert has predicted that if an attack would come from outer space humanity won't be able to repel it. This prediction is clearly unfolding before our eyes, since humans is focused too much on defending its countries from other enemy countries, while not paying attention to the possibility of an invasion from outer space. As Robert has predicted humans will be able to be an inter planetary species, it is only a matter of time.

**References:**

1. Robert A. Heinlein. (2015, June 11). Retrieved July 16, 2020, from https://www.bookseriesinorder.com/robert-a-heinlein/

2. Https://www.fantasticfiction.com, W. (n.d.). Robert Heinlein. Retrieved July 16, 2020, from https://www.fantasticfiction.com/h/robert-heinlein/

3. 19 Predictions For The Future From Robert A. Heinlein. (2019, November 30). Retrieved July 16, 2020, from https://www.writerswrite.co.za/robert-a-heinleins-19-predictions-for-the-future/

# Robert Heinlein and the end of Communism

BY: Svetozar Zirnov, Austin Mardon, Peter Johnson, John Johnson, and Elisia Snyder

Robert Heinlein, the great science fiction writer, and the greatest of his kind, has predicted that the ideology known to us as communism will vanish from the earth. Robert has been dubbed to be the one who ushered in "The golden age of science fiction". Many of Robert's predictions came true, and others that did not are starting to come true. The ideology of Communism, has been founded by a man from Germany, named Karl Marx. His goal was to find an alternative to the well known system of Capitalism. The system of capitalism was based mainly on the ability of the wealthy to become even wealthier, and allow for the accumulation of wealth for the 1%, and the other 99% were to work hard, in order to supply the wealth to the 1%. The system of capitalism, has in a way ushered in the exploitation of the working class, as well as taken away the social security that the working class requires. The system that Karl Marx has created has been the system of putting the interests of the 99% of the population, before the 1% of the rich. This system he has called Communism. Karl has produced the book that basically highlighted the aspects of the communist system, how it is to operate, as well as how it could be established. The book he wrote was called "The Communist Manifesto". While under capitalism, humanity is being divided into various classes, under Communism, Karl wanted to assure that there will be more equality between people, he wanted all people, from various nationalities to be equal, and to see each other as an equal to them. Communism was clearly based on "to each one according to their needs". While under capitalism, more and more businesses are being owned by the private sector, and various industries are being privatized in order to reduce government intervention in the system, under Communism everything belonged to the state, and thus was given to each one according to their needs, there was no competition between companies, and that is why the quality of the products made was way superior to the ones produced under the competitive capitalist system. Karl created a system where the problem of homelessness was entirely solved. That sounded better to many people than did, the inequality and the degradation that has been experienced by people living under capitalist regimes. That lead to the popularity of Communism in the 20th century. Many countries have taken the revolutionary step to establish a Communist system within their countries. Some examples include, but are not limited to the Soviet Union, Cuba, Venezuela, Bolivia, North Korea, Laos, China, and Vietnam, among others. While Communism became popular, also did Fascism become popular. Many countries, such as Germany, Italy, Spain, and Bulgaria, among others, had Fascist regimes. Fascism was also called corporationism by Benito Mussolini, the founder of Fascism, and an Italian dictator. By that he meant that under Fascism everything was privatized and belonged not to the state like in Communism, but to corporations. Communism was also revolutionary in that, the Soviet Union, was of the best examples of Communist rule, was able to conquer the German forces of Nazism, and win world war two. Nazism, is an ideology which is a derivative of Fascism, it is Fascism that is also very racist in nature. Communism was quite popular until 1991, when the country known to us as the Soviet Union has broken up into various smaller countries. Most of the countries that were comprising the Soviet Union have transitioned to the Capitalist system, while Belarus, one of the countries that constituted the Soviet Union, has remained under a dictatorial Communist rule until current times. Slowly, around the world the ideology of Communism has been growing to become unpopular, as many countries were swinging from the far-left of the political spectrum, further to the political right. Some countries didn't want to fully give up on the various benefits that they inherited under the Communist rule, and thus have transitioned to a system called Social Democracy. under the system of Social Democracy, there was a mixed market economy, where it was partially Capitalist, and partially Socialist. While the Social Democratic system has existed within the boundaries of Capitalism, it nevertheless have kept various aspects of Socialism, such as a national pharmacare law, which allowed for a health care system that was fully paid by the government and belonged to the state. Subsidized education was also part of that system, thus providing students from families who were not able to earn enough to study, the opportunity to be educated. Also, various other social programs, such as low-cost childcare, subsidized dental care and a guaranteed government pension plan. Thus, giving people the social security they needed while living within the boundaries of the Capitalist system. Some other countries have taken a more Liberal economic position. Liberalism was mainly based on having a fully Capitalist system, while at the same time supporting some social programs, such as subsidized health care, and low-cost childcare, among others. While the difference between Liberalism and Social Democracy, being that Social Democrats wanted to impose stricter rules on corporations, while Liberals, wanted them to be slightly less stricter. Some other countries have went full swing conservative. Conservatives were supporting the corporations, more than the other ideologues and have given them the best incentives. They have taken out all the strict rules, that the Liberals, and Social Democrats have imposed on them. Conservatives believed that poverty was inevitable and the government cannot do anything to fix it, or to make it better. Thus, the corporations and the big businesses were the biggest beneficiaries of the conservative system of aggressive Capitalism. As Robert Heinlein has clearly predicted that are now less, and less countries who still have a Communist system in place, and many of them are moving away from Communist ideals. Thus, the Capitalist system is becoming the main world economic system in which we currently live, and will most likely remain the only system in the near future.

References:

1.  Robert A. Heinlein. (2015, June 11). Retrieved July 16, 2020, from https://www.bookseriesinorder.com/robert-a-heinlein/

2. Https://www.fantasticfiction.com, W. (n.d.). Robert Heinlein. Retrieved July 16, 2020, from https://www.fantasticfiction.com/h/robert-heinlein/

3. 19 Predictions For The Future From Robert A. Heinlein. (2019, November 30). Retrieved July 16, 2020, from https://www.writerswrite.co.za/robert-a-heinleins-19-predictions-for-the-future/

# Robert Heinlein and World Poverty

BY: Svetozar Zirnov, Austin Mardon, Peter Johnson, John Johnson, and Elisia Snyder

Robert Heinlein, the great science fiction writer, and the greatest of his kind, has predicted that the ideology known to us as communism will vanish from the earth. Robert has been dubbed to be the one who ushered in "The golden age of science fiction". Many of Robert's predictions came true, and others that did not are starting to come true. Robert has clearly predicted in one of his books that the world food supply will be greatly reduced, which in turn may increase poverty around the globe. There are many issues that we are facing today in regards to the world food supplies. The greatest of those issues is the issue of climate change, it affects us in every way we could possibly imagine. A big part of the world food supply is coming from farms around the world, farmers have to grow their grow, which later on they sell, in order for the crops to be exported to the market. There are various crops that farmers grow, they may be vegetables, fruits, wheat, and grain, among others. Since, climate change is taking a harder swing here on earth, we are seeing more extreme and severe forms of weather, such as more thunderstorms, hurricanes, tornadoes, heavy winds, and extreme temperature fluctuations. All this is happening, not because it is natural, but because climate change is coming to its final and most devastating point. Many communities around the world who used to experience more cold and frosty weather during the year, are getting more warmer days, and more sunshine, while communities that have experienced hotter and dryer weather, are having more rain and humidity, and possibly even snowfalls. This results in devastating effects on the crops that farmers grow around the world, as well as the global food supply. Those fluctuations in temperatures, and increased severe weather, cause great damage to farmer's crops. As the farmers are exporting less produce to the markets, the prices of the produce also increase rapidly. Climate change may soon be affecting us in even more devastating ways, such as coastal floods, and extreme droughts, or excess rains and snow. As soon as the world economy continues to run various polluting refineries, climate change will continue to be a reality. The pollution may come from various industrial factories that humans run, such as the oil refineries, gas refineries, pulp and paper mills, steel mills, mineral and metal processing factories. Those industrial facilities release into the air, a chemical called carbon dioxide (CO2), which in turn thins the ozone layer, and makes climate change a possible reality. Another issue that humanity faces is the issue of overpopulation. Overpopulation is becoming a serious threat to the global food supply, since the more individuals there are, the more food needs to be made for them to consume, but since the amount of food in the world is limited, this may increase poverty and send even more people to the streets. Laws such as those introduced in overpopulated countries, such as China and India, might help to control the issue of overpopulation and reduce the pressure on the global food supply. Another important issue that we face in regards to the global food supply is government policies. Government policies may work in various ways, some might help to assure that all humanity is being supplied with their needs of food, and others might make all the food supply to accumulate at the hands of the wealthy, and thus while having the necessary food supply to feed all people, the people are not going to get enough food supplies, because of certain government policies. Under the system of Socialism, which was very popular in the 20th century, the motto was "to each according to their needs", thus the government was there to assure that everyone will have food supply, and that each family will get exactly the right amount of food supply, that that family requires. This is almost ideal, given the fact that the global food supply is limited, thus there will not be an accumulation of food supply in the hands of the wealthy, and thus reduce poverty, and possibly even drive it out of existence. Under the Capitalist system, which is prevalent in our days, the opposite is true, Capitalism believes that poverty is inevitable, thus does not control the amount of food that is flowing to each family, thus contributing to the disturbed notion that the wealthy may accumulate a large portion of the food supply, while those at the bottom of the wealth pyramid, may not even have enough food supply to survive. Thus, this system is contributing to poverty, and during economic recessions, to even mass poverty. Under the Capitalist system, poverty is not only possible, but is almost inevitable, since the amount of food supply that a specific family can own, does not depend on their needs, as under the Socialist system, but depends on the amount of wealth they can earn and accumulate. Thus, the wealthier you are, the larger is the food supply that you can own. The limits of the global food supply has even been overseen in the 20th century, where the US government wanted to produce larger amounts of food that it used to produce in the past, thus it began the use of GMO's (Genetically Modified Organisms). While using GMO's has contributed to the US government being able to produce larger amounts of food, then it did in the past, it has also contributed to various illnesses, and allergies, within the human population. For example, while in the past, milk was produced entirely on natural basis, thus the cow was consuming natural hay, nowadays, cows in various farms around the world are consuming hay the contains GMO particles in it, thus we can already come to the conclusion that the milk that the cow will produce, will not be entirely natural, but rather contain GMO particles in it. Thus, while more food is being produced the quality of food is going down rapidly. As the world faces the various issues of climate change, overpopulation, and poor government policies and regulations, the world food supply may rapidly decrease, and thus contribute to, as Robert predicted mass poverty around the world.

References:

1.  Robert A. Heinlein. (2015, June 11). Retrieved July 16, 2020, from https://www.bookseriesinorder.com/robert-a-heinlein/

2. Https://www.fantasticfiction.com, W. (n.d.). Robert Heinlein. Retrieved July 16, 2020, from https://www.fantasticfiction.com/h/robert-heinlein/

3. 19 Predictions For The Future From Robert A. Heinlein. (2019, November 30). Retrieved July 16, 2020, from https://www. writerswrite.co.za/robert-a-heinleins-19-predictions-for-the-future/

Catherine and Austin Mardon meet Pope Francis.
(November 6, 2019)